# NANOSTRUCTURED MATERIALS: MODELING AND SIMULATION CONCEPTS

# NANOSTRUCTURED MATERIALS: MODELING AND SIMULATION CONCEPTS

*By*

*Dr. Alan Parachoniak*

2016

SBS Publishers & Distributors Pvt. Ltd.
New Delhi

ISBN 13 : 9789380090627

First Published in 2016



Published by:

SBS PUBLISHERS & DISTRIBUTORS PVT. LTD.

2/9, Ground Floor, Ansari Road, Darya Ganj,

New Delhi - 110002,

INDIA

Tel: 0091.11.23289119 / 41563911

Email: mail@sbspublishers.com

www.sbspublishers.com

# Preface

Nanostructured Materials are a new class of materials which provide one of the greatest potentials for improving performance and extended capabilities of products in a number of industrial sectors, including the aerospace, tooling, automotive, electric motor, duplication, and refrigeration industries. Encompassed by this class of materials are multilayers, nanocrystalline materials and nanocomposites. Their uniqueness is due partially to the very large percentage of atoms at interfaces and partially to quantum confinement effects. Nanostructured materials may be defined as those materials whose structural elements - clusters, crystallites or molecules - have dimensions in the 1 to 100 nm range. The explosion in both academic and industrial interest in these materials over the past decade arises from the remarkable variations in fundamental electrical, optical and magnetic properties that occur as one progresses from an `infinitely extended' solid to a particle of material consisting of a countable number of atoms. This details recent advances in the synthesis and investigation of functional nanostructured materials, focusing on the novel size-dependent physics and chemistry that result when electrons are confined within nanoscale semiconductor and metal clusters and colloids. Carbon-based nanomaterials and nanostructures including fullerenes and nanotubes play an increasingly pervasive role in nanomaterials science and technology and are thus described in some depth. In this book current Modeling and Simulation Concepts and the future prospects for nanostructured materials are discussed. Nanostructured Materials: Modeling and Simulation Concepts is an extensive update to the exceptional concept of this rapidly advancing field. It covers the important synthesis and processing methods for the production of nanocrystalline materials and selected properties of nanostructured materials. Potential or existing applications are described as appropriate throughout the book.

**Editor**

# Contents

# Chapter 1

# IMPACT OF PARAMETER VARIATION IN FABRICATION OF NANOSTRUCTURE BY ATOMIC FORCE MICROSCOPY NANOLITHOGRAPHY

Arash Dehzangi[1 *], Farhad Larki[1], Sabar D. Hutagalung[2*], Mahmood Goodarz Naseri[3,4], Burhanuddin Y. Majlis[1], Manizheh Navasery[3], Norihan Abdul Hamid[1], Mimiwaty Mohd Noor[1]

[1]Institute of Microengineering and Nanoelectronics, Universiti Kebangsaan Malaysia, Bangi, Selangor, Malaysia

[2]School of Materials and Mineral Resources Engineering, Universiti Sains Malaysia, NibongTebal, Penang, Malaysia

[3]Department of Physics, Faculty of science, Universiti Putra Malaysia, Serdang, Selangor, Malaysia

[4]Department of Physics, Faculty of Science, Malayer University, Malayer, Hamedan, Iran

## ABSTRACT

In this letter, we investigate the fabrication of Silicon nanostructure patterned on lightly doped ($10^{15}$ $cm^{-3}$) p-type silicon-on-insulator by atomic force microscope nanolithography technique. The local anodic oxidation followed by two wet etching steps, potassium hydroxide etching for silicon removal and hydrofluoric etching for oxide removal, are implemented to reach the structures. The impact of contributing parameters in oxidation such as tip materials, applying voltage on the tip, relative humidity and exposure time are studied. The effect of the etchant concentration (10% to 30% wt) of potassium hydroxide and its mixture with isopropyl alcohol (10%vol. IPA ) at different temperatures on silicon surface are expressed. For different KOH concentrations, the effect of etching with the IPA admixture and the effect of the immersing time in the etching process on the structure are investigated. The etching processes are accurately optimized by 30%wt. KOH +10%vol. IPA in appropriate time, temperature, and humidity.

## INTRODUCTION

The conventional techniques for semiconductor fabrication are designed in various lithographical methods based on the top-down approach. These techniques are highly developed and optimized. However, their high operating cost, multiple-step processes, and poor accessibility for nanofabrication regularly limit the applicability of these techniques. Unconventional methods, on the other hand, can provide simpler and cheaper courses to the fabrication of nanostructures. Recently, novel methods such as nano-imprint lithography (NIL) [1], [2], soft lithography [3], and atomic force microscopy (AFM) nanolithography [4], [5] have been emerged as flexible alternatives for nanoscale patterning and fabrication. These novel methods have the potential to be as future low-cost techniques in nanoscale pattern formation and replication. Among these techniques, AFM nanolithography techniques can be capable of fabricating or characterizing microarrays [6], nanoscale structures, sensors or devices.

Generally, AFM based techniques can be categorized into three main groups in terms of their operational principles as force-assisted, bias-assisted [7] and thermal-assisted [8] AFM nanolithography. Tip induced local anodic oxidation (LAO), as a bias-assisted method, is one of the widely investigated processes for AFM nanolithography.

Local oxidation of silicon surfaces by using AFM lithography techniques in the air, was observed in 1994 by Snow and Campbell[9] for the first time, when they realized that it had a numerous advantage over the previous scanning tunneling microscope (STM) based method [10]. Proper choice and preparation of a sample are of great importance in LAO. Although, the first experiments started on a Si (111) surface, a hydrogen passivated p-type Si (100) was subjected to many works for LAO [11],[12]. Ionica et al. [13] have remarkably reported the electrical characteristics of the devices made by AFM nanolithography. In addition, some recent works were performed to improve the method of AFM nanolithography [14], [15].

The widely used wet etching is still an applicable technique in the fabrication of Si nanostructures. Tetramethyl-amino-hydroxide (TMAH) and potassium hydroxide (KOH) saturated with Isopropyl alcohol (IPA) are two well-known etchants, applicable for etching nanostructures fabricated by LAO technique. We reported the electrical property and fabrication of the p-type side gate nanowire junctionless transistor (JLT) by AFM - LAO nanolithography, on p-type (100) silicon on insulator (SOI), using KOH+IPA wet etching [16]. The numerical performance of the devices, regarding to the particular method of fabrication (AFM nanolithography), were also investigated for the single gate [17] and double gate structure [18]. The structure and particularly nanowire are defined by well precise crystalline plains where the etching is supposed to be stopped due to the anisotropic etching. The regularity and controllability of the cross section of the nanowire (trapezoidal cross section) after etching can be an important advantage for the fabrication of ultra-small and long nanostructures [19].

In this work, we present an elaborate investigation of experimental parameter variation impact on the fabrication of nanodevices by AFM nanolithography. To complete the previous works as a guide for future experiments, comprehensive information and a precise view of every step of bias-assisted AFM nanolithography are provided in details. Meanwhile, we will review some of the previous works to support or make comparisons. The important contributing parameters in the fabrication process, from preparation of the SOI wafer to the final extraction of the structure are presented as well. In this matter, the effect of tip materials, applying voltage on the tip, relative humidity (RH%) and the exposure time in the LAO process, as well as the effect of KOH concentration and immersing time on the SOI surface and the quality of etched structure, are investigated.

## METHODOLOGY

A nanostructure was fabricated via AFM nanolithography by scanning probe microscope machine (SPI3800N/4000, SII Nanotechnology Inc., Chiba-shi, Chiba, Japan). Local anodic oxidation was carried out on a lightly doped ($10^{15}$ $cm^{-3}$) p-type B-doped (100) SOI wafer (prepared by Unibond™ (Unibond International Ltd., Uxbridge, Middlesex, UK)), with a top silicon of 100 nm and a 145 nm buried oxide (BOX) thickness with a resistivity of 13.5–22.5 Ω cm (Soitech) [20]. The LAO procedure was performed in the contact mode, employing AFM conductive-coated tips with the force constant of 0.2 N $m^{-1}$ and the resonance frequency of 13 kHz. The important parameters for the scan console during the AFM scanning in different scan area are illustrated in Table 1.

Table 1. The outline of the scan parameters for the AFM scanning.

doi:10.1371/journal.pone.0065409.t001

| Scan area (nm) | 200 | 500 | 2000 | 1000 |
|---|---|---|---|---|
| Scan speed (Hz) | 6 | 4 | 2 | 1 |
| I Gain | 0.15 | 0.20 | 0.30 | 0.40 |
| P Gain | 0.05 | 0.10 | 0.15 | 0.20 |

doi:10.1371/journal.pone.0065409.t001

Room humidity (RH) was controllable from 50% to 80% with an accuracy of 1%. After the oxidation procedure, two steps of the wet etching process were applied to extract the nanostructure. In this study, the $SiO_2$ formed by LAO is considered as a mask on the top of the SOI surface. The unoxidized Si is the undesired area, which is supposed to be removed by KOH solution (KOH pellets of 56.11 g/mol supplied by System) as the etchant. The BOX layer was used as the etch-stop in the wet - etching process. All KOH etching experiments were carried out in a closed glass beaker with a constant temperature bath. The oxide removal is the second step in the etching process and the last step in our fabrication method. The etchant is hydrofluoric acid (HF, 49% concentration supplied by JT Baker) in aqueous solution with the well-known ability to dissolve $SiO_2$. During HF etching, the oxide mask and the native oxide were removed.

The whole process of fabrication can be divided into three separate and major steps, which are sample preparation, oxide growth mechanism on the desired area (mask) and etching processes (schematically illustrated in Fig. 1). Each step has important contributing parameters and a significant impact on the fabrication of the final structure, which will be addressed in the next section. The shape of the device consists of square pads as the contacts (source, drain and gate) and a nanowire between the square pads as the channel [16]. The morphology of the etched surfaces was also observed by using scanning electron microscopes (JEOL JSM-6360).

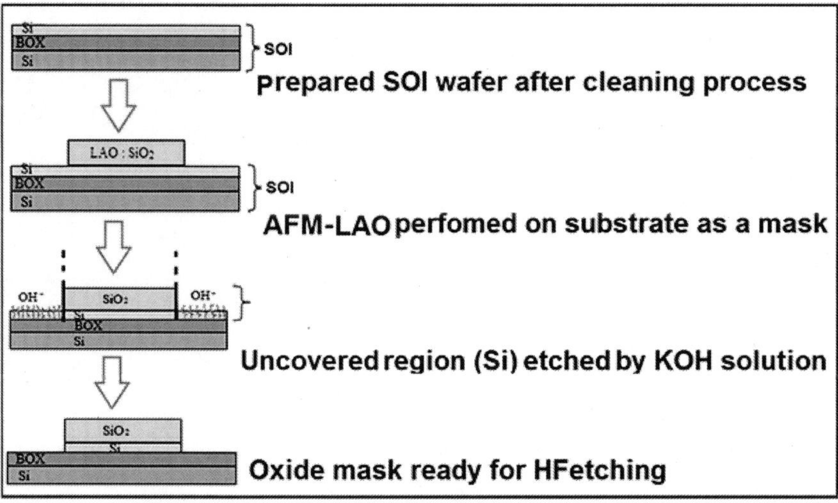


Figure 1. Schematic steps of Fabrication by AFM-LAO nanolithography.

doi:10.1371/journal.pone.0065409.g001

## RESULTS AND DISCUSSION

### Cleaning and Sample Preparation

The sample preparation is an important part in nanofabrication. A silicon surface in air ambient is covered by a variety of contaminations, such as organic and inorganic contaminations as well as native oxide. In order to obtain a nanometer flat oxide-resistant passivated surface for LAO, it is necessary to provide contaminant elimination, native oxide removal, and silicon dangling bonds saturation by hydrogen atoms [11]. Werner Kern developed one of the standard cleaning procedures in industry in 1965, while he was working at the Radio Corporation of America (RCA). It is known as the RCA method [21]. The RCA cleaning process is implemented in sample preparation as a reliable method. Prior to RCA, the samples are exposed to ultrasonic energy in an ultrasonic bath. The RCA method was modified and optimized to extract the best possible output, especially for the SOI sample with a thin layer of Si. The modified RCA applied to the samples is summarized in Table 2 in three separate steps. During the optimization of the cleaning process, major parameters were the ratio of chemical concentration with deionized water (DIW), temperature, and time.

Table 2. Steps of the modified RCA method used in the preparation process.

| Step1 | Step2 | Step 3 |
|---|---|---|
| Solution of DIW: $H_2O_2$: $NH_4OH$, (5:1:1) was prepared in the glass beaker and heated to 77 to 80°C | Solution of 6:1:1 DIW:$H_2O_2$:HCl, was prepared in the glass beaker and heated up to 77 to 80°C | Solution of HF: DIW with a ratio of 1:100 was prepared in the polypropylene beaker |
| The wafers were soaked for 12 minutes and the temperature was maintained not more than 80°C | The wafers were soaked for 12 minutes at temperature not more than 80°C | The wafers were soaked in the solution for 20 −30 seconds, until the wafer became hydrophobic |
| The wafer was rinsed with DIW for 1 minute | The wafer was rinsed in DIW for 1 minute | The wafer was rinsed in DIW for 1 minute and immersed for 2–3 minutes |

doi:10.1371/journal.pone.0065409.t002

doi:10.1371/journal.pone.0065409.t002

In the first step of the cleaning process, the ammonia/peroxide mixture (APM) oxidizes the silicon surface using hydrogen peroxide ($H_2O_2$), and simultaneously removes the silicon oxide by means of ammonium Hydroxide ($NH_4OH$, concentration of 25%wt) at alkaline pH [22]. In the second step of cleaning to remove ionic contamination by using hydrochloric/peroxide mixture (HPM), metal ions and metal hydroxides will be solubilized. Hydrochloric acid (HCl) generates a metal cation in solution that forms an ionic bond with the chloride anion in order to rinse all metal tracks off the silicon surface [23].

In this step, temperature and time interval control play a crucial role in the nanometer flatness of the silicon surface. Fig. 2 shows the effect of improper temperature and time interval on Si surface. For a solution with longer time interval (25 min), the surface roughness increases (Fig. 2a) and for a solution with a higher temperature (>100°C), the surface flatness degrades (Fig. 2.b). Any change in temperature more than 3°C during the cleaning process (when the temperature exceeds 80°C) would give rise to inequalities or even twists on the sample surface.

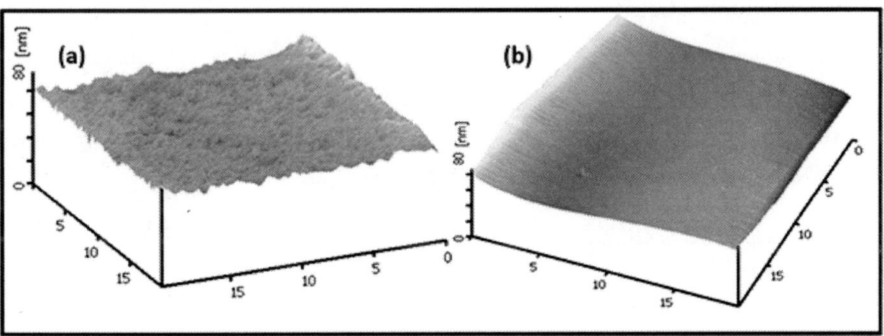


**Figure 2.** AFM images showing the effect of improper cleaning parameters on the SOI surface (a) long time interval (25 min), and (b) high temperature (>100 °C).

doi:10.1371/journal.pone.0065409.g002

To remove the native oxide and saturate the silicon dangling bonds, concerning (100) silicon surface, an aqueous HF (concentration of 49%) solution was employed. Fig. 3(a–d)demonstrate the impact of HF concentration variation, from 0.25% to 1.0% in DIW (in the steps of 0.25%) at the temperature of 242°C and humidity of 672%.

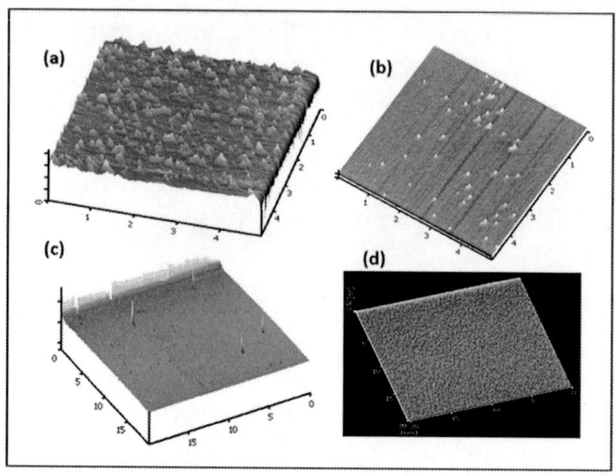


Figure 3. The effect of HF concentration on native oxide removal during the sample preparation.

doi:10.1371/journal.pone.0065409.g003

It was also found that the type and brand of SOI wafer can make some differences in the details of each step of the cleaning process. Fig. 4 shows the AFM images of the SOI arbitrary sample before and after the optimized cleaning process.

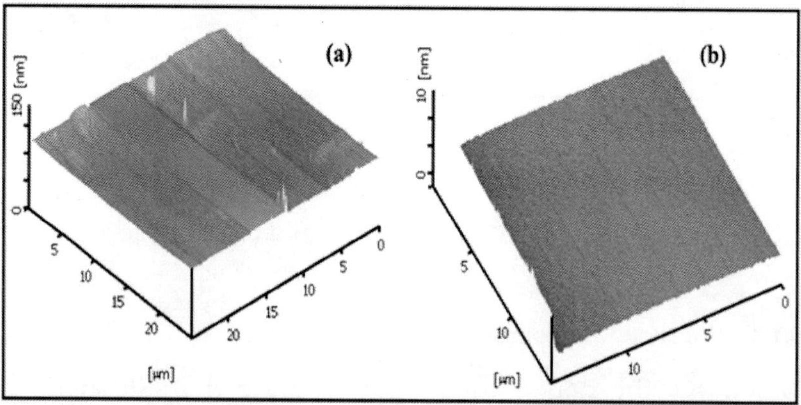


Figure 4. AFM images of the SOI substrate (a) before and b) after optimized RCA.

doi:10.1371/journal.pone.0065409.g004

## LAO- AFM

After the cleaning process, Si-H bonds would be formed on the surface of the SOI. When the surface was de-passivated, the first layer of Si-Si bonds became polarized due to the high electron negativity of the $OH^-$. During LAO, a biased AFM tip (−5 to −15 V) is operating under an ambient humidity to locally oxidize the surface of a desired sample. For few nanometers gap between the tip and the surface, AFM tip bias would generate a field of $10^8$ V/m to $10^{10}$ V/m. The available water meniscuses, as a water bridge in the gap, are dissociated by this negative tip bias. Fig. 5 schematically shows the LAO-AFM process.

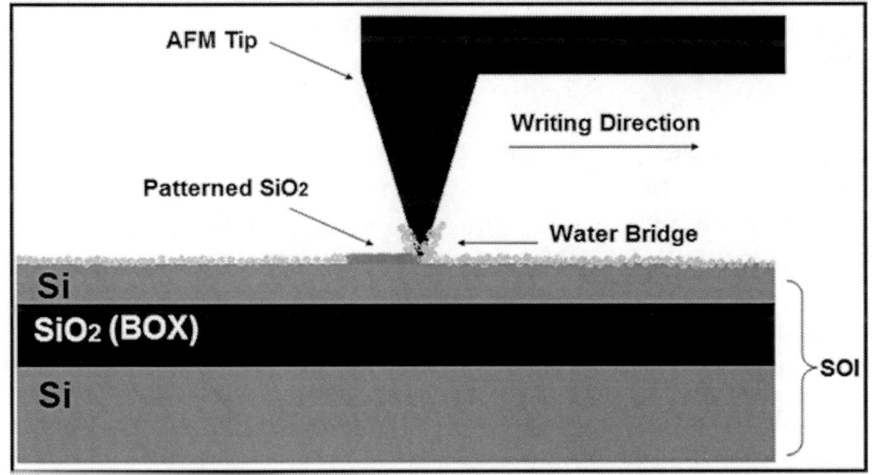


**Figure 5. Schematic of AFM – LAO process on SOI substrate.**

doi:10.1371/journal.pone.0065409.g005

The oxide growth mechanism has been elaborately described in the literature [24], [25]. Complete understanding or quantifying process is complicated due to the specific spatial details of the tip/surface contact profile, as well as the cantilever motion under the applied bias[26]. However, some explanations and models are proposed to explain the anodic oxidation mechanism. In Table 3, some of the basic proposed models for the local oxidation by the AFM nanolithography are summarized. All the research groups agreed that the oxidation mechanism and kinetics are closely related to the electric field, surface stress, water meniscus formation, and $OH^-$diffusion [27], [28], [29], [30].

Table 3. Various AFM local anodic oxidation models.

doi:10.1371/journal.pone.0065409.t003

| Oxidation model | Description |
|---|---|
| **Cebrera-Mott model** | The thickness of the oxide is governed by a diffusion limited of the electric field [31,64] |
| **Power- law model** | The AFM oxidation is only observed for a voltage which exceeds more than a doping dependent threshold, above which the oxidation kinetics follows a power law [32] |
| **Log kinetic model** | The height of the oxide is proportional to log(1/t) and the linear behavior between 1/h vs log v [36] |
| **Space charge model** | i) Varied space-charge dependence of oxidation is a function of substrate doping type/level. |
| | ii) Space- charge effects are consisted of the rapid decline of the high initial growth rate. |
| | iii) The Alberty- Miller scheme describe a direct pathway for reaction of oxyanions with silicon at the $Si/SiO_2$ interface, and an indirect reaction pathway, mediated by the trapping charge defects located at the interface [44,65]. |

doi:10.1371/journal.pone.0065409.t003

Several models have been proposed for explaining the oxide growth on the surface of semiconductors such as Cabrera–Mott model [31], power-law model [32], direct log kinetic model [33] and space charge model [34]. Marchi et al. [35] reported a significant increase of the oxide growth by an ozone supply in ambient air which implies that the oxide growth process is limited both thermodynamically and kinetically.

The Stievenard model [36], which is modified from the Cabrera-Mott model, is the best match with our experimental results, regarding to the quantitative analysis of the measured oxide height. Based on this model, the ionic diffusion of $OH^-$ and $O^-$ takes place through the oxide layer, from the oxide surface towards the $Si/SiO_2$ interface. The applied electric field enhances the diffusion mechanism, due to the Mott potential difference between the initial oxidized silicon and the adsorbed oxygen layer [36]. Under a particular condition, which is extendable into our case, the height of the oxide thickness varies linearly with the applied voltage.

## AFM tip

In the LAO process, in addition to the major factors like RH%, applying voltage, exposure time or preparation process, some minor factors exist, which are able to make a significant effect in details. The type of the AFM tip can be very important since the coated layer of a tip enhances the laser reflectivity of the cantilever. The tip should be specified for the contact mode (tapping mode and non-contact mode tips are nonfunctional in this study). It is worth mentioning that for patterning very small oxide layer on the top of the SOI surface, some remarkable work has been performed by using the tapping mode or non-contact mode [37]. We examined three types of the contact mode tips, as the gold coated (Budgetsensors Au coated ContGB-G), the Chromium/Platinum coated conductive probe (Budgetsensors Cr/PtcoatedCont-G) and the Alumina reflex coated (Budgetsensors ContAl), among which the last one was not applicable.Table 4 illustrates the optimized parameter comparison during the LAO process for two different AFM tips. Both tips had the force constant ($k_c$) of 0.2 N $m^{-1}$ and the resonance frequency of 13 KHz (tip radius <25 nm).

Table 4. LAO Parameter comparison for Au and Cr/Pt coated AFM tip.

doi:10.1371/journal.pone.0065409.t004

| Type of AFM - tip | Applied voltage | Writing speed | Scan speed | Force reference |
|---|---|---|---|---|
| Conductive Au-coated | 8 V | 0.6 μm/s | 1.0 μm/s | −0.2 N |
| Conductive Cr/Pt coated | 9V | 0.5 μm/s | 1.0 μm/s | −0.1 N |

doi:10.1371/journal.pone.0065409.t004

Applying Voltage at AFM Tip and Exposure Time (Writing Speed)

The writing speed and the AFM tip voltage are two major parameters to optimize the LAO process. A decrease was observed in the oxide height by increasing the writing speed. At the same humidity and temperature for Au coated tip, the width comparison of the oxidation tracks for different writing speeds is shown in Fig. 6c at RH 68% and 23°C with a constant applied voltage (8V).

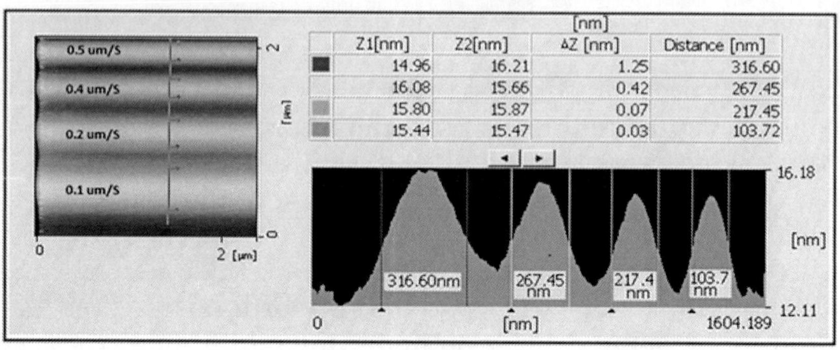


Figure 6. AFM profile image for width comparison of the oxide track for different writing speeds during LAO (Au coated tip at 8V applied voltage).

doi:10.1371/journal.pone.0065409.g006

Low writing speeds made wider oxidation tracks on the substrate, whereas faster writing speed might provide negligible track for constant applied voltage. It can be explained by the fact that as the writing speed increases, the numbers of diffusible ions into the Si/$SiO_2$ interface decreases. For Cr/Pt coated tip with the writing speeds faster than 1.0 µm/s, the oxidation tracks were not recognizable. It was observed that in comparison with the Cr/Pt coated tip at the same applied voltage, the Au coated tip provided thicker oxide tracks for different writing speeds. Measured thicknesses at different writing speeds for two types of AFM tip are summarized in Table 5.

Table 5. Oxide thickness for different AFM tips (RH 68%, T = 23°C).

doi:10.1371/journal.pone.0065409.t005

| **Writing speed** | **2.0 µm/s** | **1.5 µm/s** | **1.0 µm/s** | **0.8 µm/s** | **0.5 µm/s** | **0.3 µm/s** |
|---|---|---|---|---|---|---|
| **Oxide thickness for Au-coated tip** | ~ 0 | 1.3 nm | 2.4 nm | 3.1 nm | 3.2 nm | 3.4 nm |
| **Oxide thickness for Cr/Pt coated tip** | ~ 0 | 1.0 nm | 2.0 nm | 2.5 nm | 2.8 nm | 3.0 nm |

doi:10.1371/journal.pone.0065409.t005

Regarding to the particular condition of the experiment, the best results were extracted by Cr/Pt coated conductive probe, since the quality of the oxide layer, in terms of uniformity or narrowness, was more applicable for the purpose of fabrication (especially after etching process). Au coated conductive probe provided a thicker oxide layer after oxidation, though with a lower oxide quality. A further comparison of the impact of different AFM tips on the oxide thickness will be addressed later (Fig. 7).

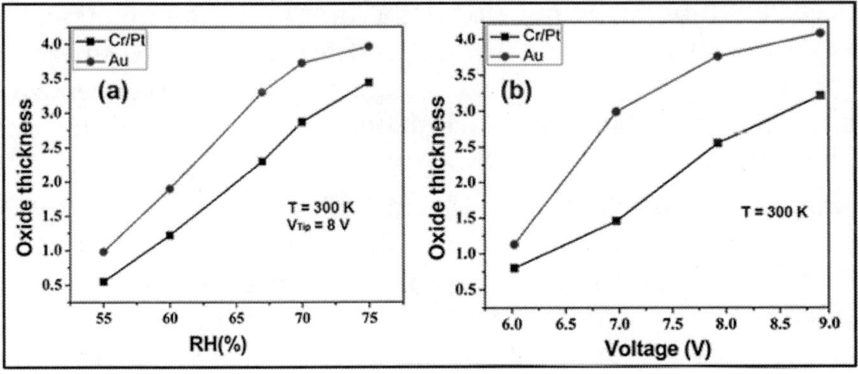


Figure 7. Relation of the oxide thickness (nm) with a) RH %, and b) applied voltage (68% of the RH%), for different AFM tips and the writing speed of 0.6 μm/s.

doi:10.1371/journal.pone.0065409.g007

As mentioned before, the oxidation rate in AFM - LAO is directly proportional to the field strength and applying voltage on the tip. At a given writing speed, we observed that the oxide thickness is enlarged with the increase in the applied voltage. Interestingly, by increasing the applied voltage on the AFM tip, the width of the oxide tracks does not show any significant dependence which is in agreement with previous reports [38].

## Relative Humidity

Another parameter that plays a vigorous role in LAO process is RH%, which is due to the necessity of water molecules in an ambient air during the oxidation process. The oxidation rate is sensitive to humidity and the oxide thickness increases at high RH%. The impact of humidity on the oxidation procedure was previously reported

[39]. For two different types of the AFM tip, at constant applied voltage (8V) and writing speed (0.6 μm/s), the relation between oxide thicknesses and RH% is shown in Fig. 7a (at room temperature). It can be seen that the oxide thickness increases as a linear function of RH%. Presence of the humidity around the electrodes influences the oxide growth significantly [24], [40] (the water bridge). In fact, high concentration of oxyanions under the tip during the initial state of oxidation, as well as lateral oxyanions diffusion on the Si surface are important factors affecting the oxide growth [41]. The concentration of this oxyanions on the surface depends on the size of the water bridge and the electric field strength.

Fig. 7b indicates that the oxide thickness depends on the applied voltage for constant writing speed (0.6 μm/s) and RH% (68 RH%). As it can be seen, Au coated tip provides a thicker oxide layer for all applied voltages in comparison with the Cr/Pt coated tip, which is also applicable for different RH% (Fig. 7a). Since the oxide growth on Si surface was not significant below 5 V, this voltage can be considered as the threshold voltage for oxidation, when the writing speed is 0.6 μm/s.

Humidity and temperature drastically affect the device structure during the patterning process. Normally to get a good pattern, the range of the temperature variation must be ≤4°C and the range of humidity variation must be ≤4%. For both Cr/Pt and Au coated tips, the observation reveals that the best results were in the range of 22–26°C for the temperature and 65–68% for the RH%. Fig. 8 shows the AFM topography and profile image of the arbitrary sample patterned by AFM –LAO, when the optimized parameters for Cr/Pt AFM tip (Table 3) are applied. The patterned structure is well-shaped and the thickness of 3 nm is an acceptable range for the oxide layer to act as a mask.

It is important to mention that applying high voltage may provide a thicker oxidation on the Si surface, but the pattern will be wider with lower uniformity. This can be improper for the purpose of nanodevice fabrication in order to extract the narrower nanowire after the etching process. Furthermore, for a high applied voltage in a high RH%, the shadow effect around the pattern will be inevitable.

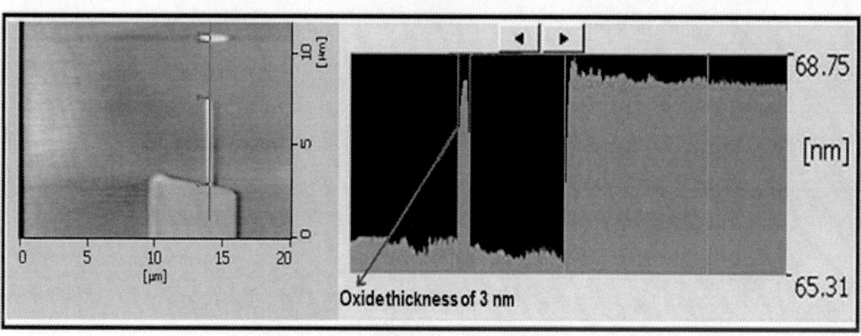


Figure 8. AFM images of the LAO patterned sample with the optimized parameter (Cr/Pt tip).

doi:10.1371/journal.pone.0065409.g008

## Shadow Effect

In the LAO process, the diffusion rate is proportional to the electric field strength. Larger applied voltage induces a larger electric field which causes a stronger interaction between the tip and the surface [41]. At high RH%, the electrochemical interaction between the tip and the surface becomes much stronger which prevents the effective oxidation at the center of the growing oxide. In this condition, the current expands horizontally through the surface layer, leading to the creation of shadows around the oxide pattern. According to our observation, when RH% was increased beyond 70%, the rate of increasing oxide thickness was reduced, and the shadow effect was observed. This supports the previous studies on AFM-LAO about the influence of room humidity in oxide growth on the Si surface, when the other parameters are constant [24], [41]. Fig. 9a presents the AFM topography image of the pattern with shadow effect around the square pads prepared by the LAO technique at high RH%. The AFM profile image for the same structure (Fig. 9b) presents the oxide thickness comparison for the pattern and shadow area around it. It shows that for the thickness of 3.12 nm for the oxide thickness, the shadow thickness is 1.38 nm around the oxide pattern.

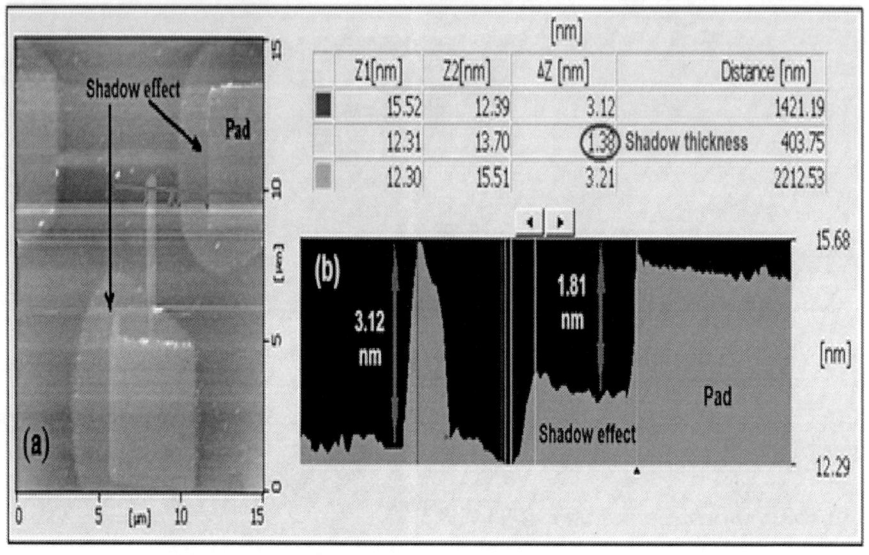


Figure 9. a) AFM image of the pattern after LAO in high RH% with shadow effect, b) profile view of the same sample showing the oxide and shadow thickness.

doi:10.1371/journal.pone.0065409.g009

## KOH Etching Process

Different research groups reported different wet and dry etching materials and processes. For instance, Hydrazine is one of the etchants with the advantage of negligible degradation during deep etching [10]; however, extreme caution is needed since it is highly toxic and potentially explosive. Dry plasma etching method, e.g. RIE method, is also employed. It allows obtaining a great variety of etching profiles ranging from isotropic etch to vertical sidewalls. However, the selectivity between the oxide mask and the silicon is generally poorer for liquid etchants [42].

The properties of non-annealed oxide created by AFM-LAO are different from thermally grown oxide, with a lower density and higher dielectric constant of 5.2 vs 3.9 [43]. Non-annealed oxide has even a greater $H_2O$ content by weight [44] in comparison with the thermal case. According to the low quality of the grown oxide by

LAO, after the etching procedure, the selectivity of the agent is a critical parameter to achieve a high aspect ratio for nanostructures.

KOH saturated with IPA is a well-known etchant in the anisotropic chemical wet etching technique, which can used for the etching of nanostructures fabricated by the LAO technique. Several research teams have elaborately investigated the mechanism of the KOH anisotropic etching [45], [46], [47].

In order to extract the final structure after oxidation, two steps of etching process were conducted. KOH wet etching is a very significant part in the fabrication of devices. Some difficulties such as ill-etched, over etching or even contamination are hardly avoidable during the wet etching process. Accordingly, accuracy and precaution are essential to be given. Etching time (immersing time of a substrate in an etchant bath) depends on the thickness of the Si layer on top of the SOI wafer, which is desired to be removed. Considering the rate of the Si removal for KOH etching as 0.28 μm/min [48], the etching time of 20–22 s was selected to remove 100 nm top Si layer.

Fig. 10(a–c) show AFM images of the etched substrates for different KOH percentage in solution (%wt) from 10% up to 30%. The solution was heated up to 63°C, stirred at 600 rpm. The substrate was immersed for 20–22 seconds and rinsed by DIW for a few seconds after being removed from the etching solution. It can be seen that the surface roughness is changed as the concentration of KOH increases up to 30%. At low KOH concentrations, the surface is rougher and the formation of insoluble precipitates can be observed. Roughness reduction of the Si surface, due to concentration increase, was reported in previous works [49], [50]. Similar observations were also reported in previous works [51], [52]. However, concentration is not the only main factor that determines the surface roughness; temperature and etching time are also contributing in the final roughness of the Si surface. In reported cases, the etching temperature was normally taken between 55°C–70°C to reach the acceptable edge cutting for patterned nanostructure.

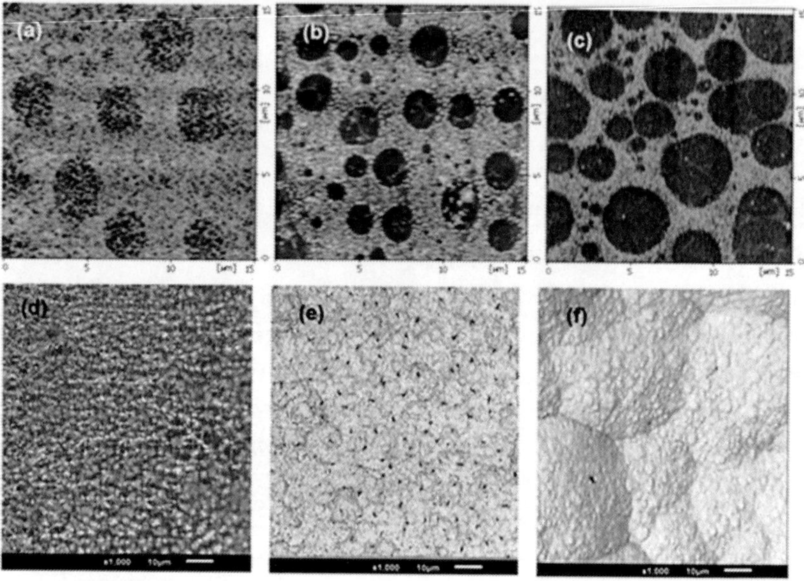

Figure 10. AFM images of the Si surface etched with KOH: (a) 10%wt (b) 20%wt (c) 30%wt, and SEM images of Si surface etched with 30%wt.

KOH +10%vol IPA at : d) 50°C (e) 60°C, (f) 80°C.

doi:10.1371/journal.pone.0065409.g010

It is approved that an admixture of isopropyl alcohol (IPA) in KOH etchant can improve the smoothness of the Si surface after etching [53]. Zubel*et al.* [54] reported the effect of the IPA concentration variation (2–12%) in the KOH etchant to enhance the smoothness of the surface. The impact of the IPA admixture in KOH etching was elaborately investigated in reported cases[45], [54]. In this work, IPA (60.10 g/mol supplied by Merck) was used as the initiator to improve surface roughness and control the etch rate [52]. It is important to mention that using IPA addition may slightly reduce the etching rate, but it definitely provides a smooth Si (100) surface in the all range of the examined concentration of KOH, specially for a high KOH ratio[55]. By controlling the temperature, in high KOH concentration with IPA addition, the etching rate can even be increased compare to the pure KOH [47]. Because of the low solubility of IPA in the KOH solution, there can be some limitation in analyzing the effect of IPA concentration on the smoothness of the etched surface.

Taking into account the natural evaporation of IPA during the etching process and its boiling point (80.37°C), the temperature during the etching process must be kept around 60°C to 80°C[45], [54], when the etching time is adjustable in accordance with the certain purpose of study. The effect of 30%wt. KOH+10% vol. IPA on quality of the etching surface for different temperatures was examined in constant etching time. SEM images of the etching effect on Si surface for different temperature (50°C to 80°C) are shown in Fig. 10 (d–f). The comparison shows smoother morphology at a lower temperature, whereby physical texture is much smaller compare to a higher temperature.

In the KOH etching process, the etching rate strongly depends on the water concentration [26]. Since water content of the solution extremely decreases at a higher level of the KOH concentration, then adding IPA can liberate water particles close to the Si surface to take part in the etching process. Meanwhile, an adsorption of IPA interrupts the access of $OH^-$ ions across the surface and reduces the oxidation rate, providing equilibrium in diffusion and oxidation of reaction products, which leads to more surface smoothness [46]. On the other hand, the decrease of the reaction rate, caused by IPA absorption, may lead to obtain the uniform etching on the Si surface.

During the KOH etching process, hydrogen bubbles are produced due to the reaction between the Si and the hydrogen ion. These bubbles increase the surface roughness [56] and act as surface masks against the reaction sites at the surface [54], [57] ("pseudo-mask" phenomenon[58]). The number and size of the hillocks during the Si etching are related to the hydrogen bubbles as well. The number of bubbles is directly proportional to the temperature, and the size of bubbles is reversely related to the etchant concentration. According to Yun [45], at temperatures below 70°C, the etch rate and concentration of KOH has the lowest influence on each other,. However, the faster etch rate can be achieved by using a high-temperature (>70°C) reaction. On the other hand, the sharpness of the etched structure is essential for nanodevices. The best sharpness after etching is achievable by keeping the temperature below 70°C and optimizing the KOH concentration. By increasing the concentration, it was observed that the density of bubbles increased but the size of the bubbles decreased. The same effect was already reported in previous studies [45]. In the range of our study, due to the low thickness of the Si layer, a high temperature may have a destructive effect on fabricated nanostructure.

During the fabrication of JLT devices, the effect of different KOH concentrations, mixed with 10%vol IPA, on the surface roughness of the devices was examined [59]. These results (not shown here) revealed that the surface roughness of the nanostrcture etched by 20%wt. KOH +10%vol IPA is higher than the one etched in 30%wt or 40%wt. KOH +10%vol. IPA. Considering the etchant solution (30%wt KOH +10%vol. IPA) for the best smoothness and sharpness, compare to the other examined etchant concentrations, the optimized range of etching temperature was extracted at 63–65°C. In this condition, compare to the case with 10%wt KOH +10%vol. IPA, the size of the bubbles is expected to be smaller. The small hydrogen bubbles leave the surface quickly, and the etching can be conducted much more efficiently by improving the sharpness of the structure. At 40%wt. KOH +10% IPA, produced hydrogen bubbles are expected to be smaller and the structure is anticipated to be sharper. However, the result was different and at this range, the sharpness and the smoothness were less than the case with 30%wt KOH. The reason can be explained by a faster Si etching at a high KOH concentration, where it can induce over etching on such structure with lower sharpness, in the range of applied temperature.

To investigate the role of the immersing time in the etching procedure for obtaining the optimum result, different etching times were examined. Fig. 11 shows the SEM image of the oxide pattern after etching. In this experiment, the sample was etched by 30%wt. KOH +10%vol. IPA solution at 63°C immersed for 35 seconds. Despite of the acceptable quality of the pads' shape, the results revealed that most parts of the nanowire were gone. Long immersing time consequently destroyed the structure of the nanowire. The same effect is reported in [45], [60].

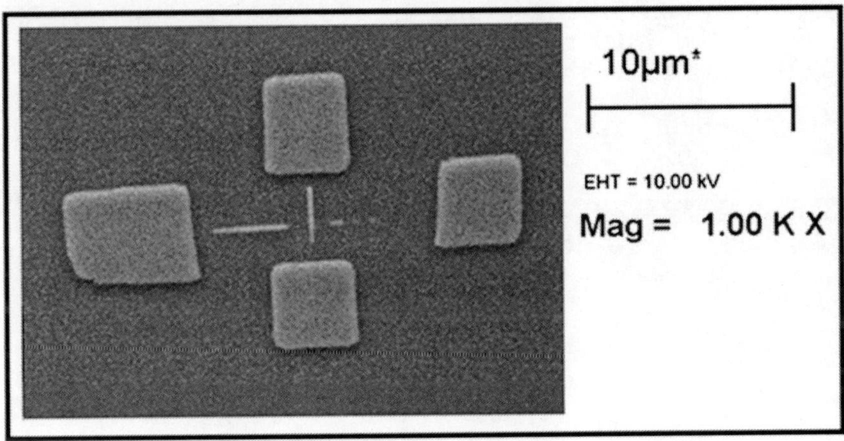


Figure 11. SEM image of the structure after etching with 30%wt.

KOH +10%vol. IPA at 63°C for 35 s.

doi:10.1371/journal.pone.0065409.g011

For shorter immersing time, the etch rate, as well as etched shape, can be different. For the solution of 30%wt. KOH +10% IPA at 63°C (optimized parameters), AFM profile images of the nanowire after etching for three different etching times are shown in Fig. 12. As it can be seen, for the shorter time of 8 s (a) and 15 s (b) compare to the longer time 22 s (c), the structures are under-etched, and the width of the nanowire are wider. This phenomenon can be related to different etch rates for the different crystallography planes. In anisotropic etching for Si (100), the etch pit is in a pyramid shape and the walls are flat and angled [45], [53]. For KOH etching, the etch rate for high index planes (e.g. (411), (311) or (211)) grows faster than the low index plane, like (110) or (111) [11], [53]. At shorter immersing time, high index planes of the Si surface would be more etched than lower index planes, and then the pit walls will be more angled. As it is expected, the etch depth for the shorter etching time, 44.71 nm for 8 s (Fig. 12a) and 73.65 nm for 15 s (Fig. 12b), are smaller than the etch depth (100.1 nm) after 22 s etching time (Fig. 12c). After 22 s of etching, the whole top Si layer (100 nm) was properly removed. Moreover, the addition of IPA to KOH solution can even more significantly reduce the etching rates of some high indexed crystal planes than the (1 0 0) plane [53].

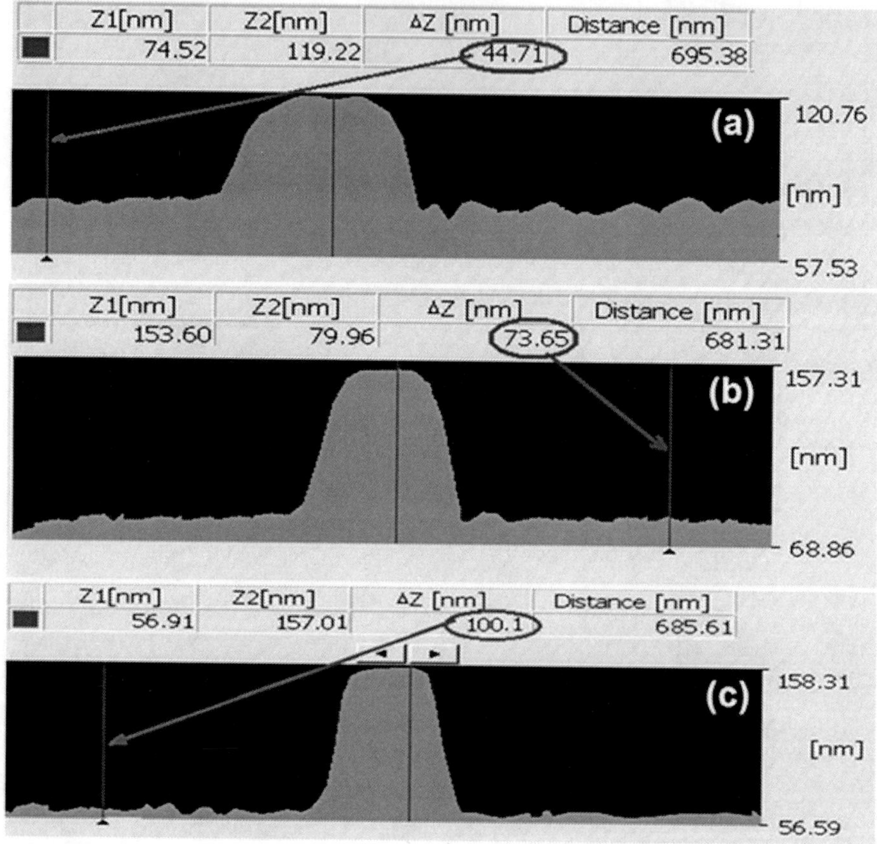


Figure 12. AFM profile images of the structure after etching for different immersing time, a) 8 s, b) 15 s, (c) 22 s (the solution of 30%wt.

**KOH +10% IPA at 63°C).**

doi:10.1371/journal.pone.0065409.g012

## Optimum Condition for KOH Etching

In this study, previous works for obtaining the optimum condition of KOH etching were considered and adopted [56], [57], [60], [61]. The best condition according to the fabrication environment and other parameters is the solution of 30%wt. KOH with 10%vol. IPA for wet etching at 63°C, 22 seconds of immersing time, stirred at 600rpm. Stirring the solution is to ensure the uniformity of the etching process. The oxide removal was performed with diluted hydrofluoric acid ($H_2O$/HF 100:1) for the time period of 16 s to 18 s. Fig. 13(a–c) show AFM image and profile of the nanostructure for an arbitrary sample after the optimized etching procedure.

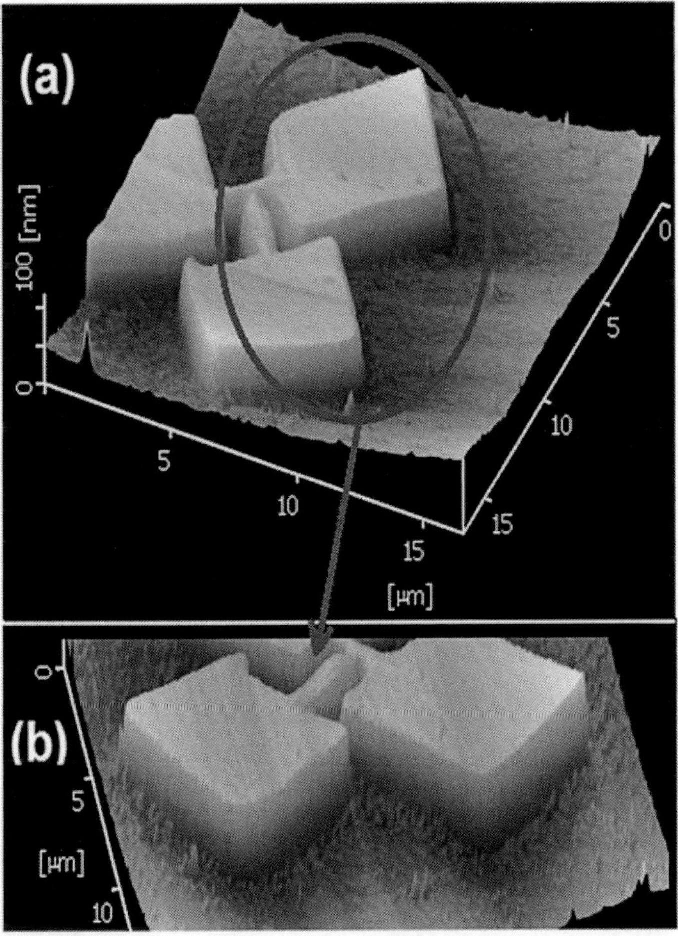


Figure 13. a) and b) AFM images of the structure after the optimized etching process.

doi:10.1371/journal.pone.0065409.g013

## Other Parameters and Issues

In the fabrication process, in addition to the major factors (mentioned in previous sections), some minor factors may be able to make significant effects in the details. For instance, the size and geometry of the tip can have different impacts on the LAO process. Some other issues can be mentioned as the order of oxidation, hidrophobicity and dielectric constant of the Si material, doping concentration or even chemical composition of the atmosphere [62], [63]. The shadow effect can significantly affect the nanostructure after etching process. Fig. 14ademonstrates the structure with shadow effect after LAO. For the same sample after the etching process, the deformation of the whole structure can be seen in Fig. 14b, where the square shape turned into the round shape.

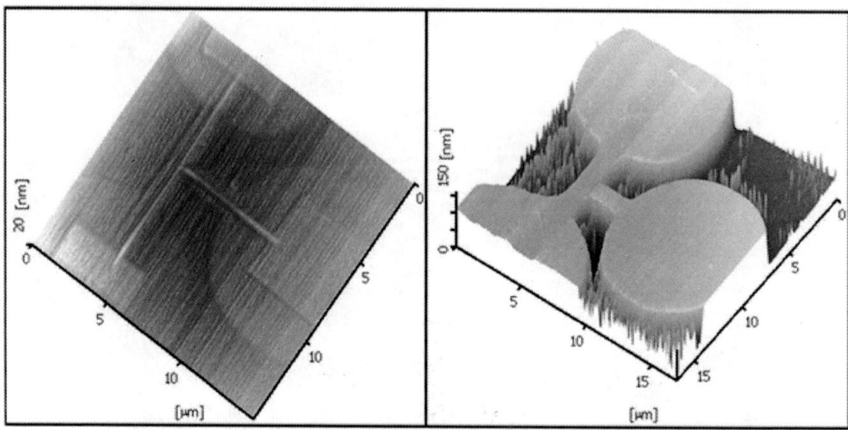


Figure 14. Shadow effect in the fabrication process on the SOI sample (a) after LAO, (b) after etching for the same sample.

doi:10.1371/journal.pone.0065409.g014

## CONCLUSIONS

In this paper, the impacts of different parameters in the fabrication of nanostructure on SOI by the improved AFM nanolithography were investigated. Sample preparation and cleaning process were modified from the RCA standard method. The impact of AFM-LAO parameters, such as the applied voltage on the tip, the relative humidity, the writing speed or the tip materials were studied to achieve the best structure. The shadow effect after the LAO process, due to the high relative humidity, was considered in this study. Two wet etching processes were implemented to extract the structure after the LAO process. The impact of the KOH concentration and KOH+IPA admixture for different temperatures on the SOI wafer were investigated. For different KOH concentrations, the effect of etching on the Si surface was also examined. The effect of the immersing time in the etching process was studied as well. The extracted optimized parameters for the etching process were 30%wt. KOH with 10%vol. IPA at 63°C, with the immersing time of 22 seconds and 600rpm stirring rate.

## ACKNOWLEDGMENTS

Authors would like to thank prof. Dr. Elias B. Saion from physics department of UPM for his great support and assistance. Authors also thank Vafa Shojamanesh and Dr. Omid Dehzangi for assisting in editing the present manuscript.

## AUTHOR CONTRIBUTIONS

Conceived and designed the experiments: AD FL SDH BYM. Performed the experiments: AD FL NAH MMN. Analyzed the data: AD FL SDH. Contributed reagents/materials/analysis tools: AD FL SDH BYM. Wrote the paper: AD FL MGN MN.

## REFERENCES

1. Guo LJ (2007) Nanoimprint lithography: methods and material requirements. Advanced Materials 19: 495–513. doi: 10.1002/adma.200600882
2. Viheriälä J, Niemi T, Kontio J, Pessa M (2011) Nanoimprint Lithography-Next Generation Nanopatterning Methods for Nanophotonics Fabrication. Recent Optical and Photonic Technologies, InTech, ISBN: 978–953.
3. Qin D, Xia Y, Whitesides GM (2010) Soft lithography for micro-and nanoscale patterning. Nature protocols 5: 491–502. doi: 10.1002/adma.200600882
4. Zhang Y, Santhanam S, Liu J, Fedder G (2010) Active CMOS-MEMS AFM-like conductive probes for field-emission assisted nano-scale fabrication; IEEE. 336–339.
5. Garcia R, Martinez RV, Martinez J (2005) Nano-chemistry and scanning probe nanolithographies. Chem Soc Rev 35: 29–38. doi: 10.1039/b501599p
6. Cook MA, Chan CK, Jorgensen P, Ketela T, So D, et al. (2008) Systematic validation and atomic force microscopy of non-covalent short oligonucleotide barcode microarrays. PloS one 3: e1546. doi: 10.1371/journal.pone.0001546
7. Xie X, Chung H, Sow C, Wee A (2006) Nanoscale materials patterning and engineering by atomic force microscopy nanolithography. Materials Science and Engineering: R: Reports 54: 1–48. doi: 10.1016/j.mser.2006.10.001
8. Knoll AW, Pires D, Coulembier O, Dubois P, Hedrick JL, et al. (2010) Probe-Based 3-D Nanolithography Using Self-Amplified Depolymerization Polymers. Advanced Materials 22: 3361–3365. doi: 10.1002/adma.200904386
9. Snow E, Campbell P (1994) Fabrication of Si nanostructures with an atomic force microscope. Applied Physics Letters 64: 1932–1934. doi: 10.1063/1.111746
10. Choi YK, Ha D, King TJ, Bokor J (2003) Investigation of gate-induced drain leakage (GIDL) current in thin body devices: single-gate ultra-thin body, symmetrical double-gate, and asymmetrical double-gate MOSFETs. Japanese journal of applied physics 42: 2073–2076. doi: 10.1063/1.111746
11. Ju C, Hesketh PJ (1992) High index plane selectivity of silicon anisotropic etching in aqueous potassium hydroxide and cesium

hydroxide. Thin solid films 215: 58–64. doi: 10.1063/1.111746

12. Graf D, Grundner M, Schulz R (1989) Reaction of water with hydrofluoric acid treated silicon (111) and (100) surfaces. Journal of Vacuum Science & Technology A: Vacuum, Surfaces, and Films 7: 808–813. doi: 10.1063/1.111746
13. Ionica I, Montes L, Ferraton S, Zimmermann J, Saminadayar L, et al. (2005) Field effect and Coulomb blockade in silicon on insulator nanostructures fabricated by atomic force microscope. Solid-State Electronics 49: 1497–1503. doi: 10.1063/1.111746
14. Pennelli G (2009) Top down fabrication of long silicon nanowire devices by means of lateral oxidation. Microelectronic engineering 86: 2139–2143.
15. Chiesa M, Cardenas PP, Otón F, Martinez J, Mas-Torrent M, et al. (2012) Detection of the Early Stage of Recombinational DNA Repair by Silicon Nanowire Transistors. Nano letters 12: 1275–1281. doi: 10.1021/nl2037547
16. Dehzangi A, Abdullah AM, Larki F, Hutagalung SD, Saion EB, et al. (2012) Electrical property comparison and charge transmission in p-type double gate and single gate junctionless accumulation transistor fabricated by AFM nanolithography. Nanoscale research letters 7: 381. doi: 10.1186/1556-276X-7-381
17. Dehzangi A, Larki F, Hutagalung S, Saion E, Abdullah A, et al. (2012) Numerical investigation and comparison with experimental characterisation of side gate p-type junctionless silicon transistor in pinch-off state. Micro & Nano Letters, IET 7: 981–985.
18. Larki F, Dehzangi A, Abedini A, Abdullah AM, Saion E, et al. (2012) Pinch-off mechanism in double-lateral-gate junctionless transistors fabricated by scanning probe microscope based lithography. Beilstein J Nanotechnol 3: 817–823. doi: 10.3762/bjnano.3.91
19. Pennelli G, Piotto M (2006) Fabrication and characterization of silicon nanowires with triangular cross section. Journal of applied physics 100: 054507–054507–054509. doi: 10.1063/1.2338599 SOITEC PTF, 38190 Bernin France.
20. Kern W, Puotinen DA (1970) Cleaning solutions based on hydrogen peroxide for use in silicon semiconductor technology. RCA rev 31: 187–206. doi: 10.1063/1.2338599
21. Kohli R, Mittal KL (2007) Developments in surface contamination and cleaning: fundamentals and applied aspects: William Andrew.
22. Verhaverbeke S, Kuppurao C, BEAUDRY C, TRUMAN JK (2002) Single-wafer, short cycle time wet clean technology. Semiconductor

international 25: 91–98. doi: 10.1063/1.2338599

23. Fang TH (2004) Mechanisms of nanooxidation of Si (100) from atomic force microscopy. Microelectronics journal 35: 701–707. doi: 10.1016/j.mejo.2004.06.022
24. Dagata J, Inoue T, Itoh J, Yokoyama H (1998) Understanding scanned probe oxidation of silicon. Applied Physics Letters 73: 271. doi: 10.1016/j.mejo.2004.06.022
25. Seidel H, Csepregi L, Heuberger A, Baumgärtel H (1990) Anisotropic etching of crystalline silicon in alkaline solutions I. Orientation dependence and behavior of passivation layers. Journal of the Electrochemical Society 137: 3612–3626. doi: 10.1016/j.mejo.2004.06.022
26. Snow E, Jernigan G, Campbell P (2000) The kinetics and mechanism of scanned probe oxidation of Si. Applied Physics Letters 76: 1782. doi: 10.1016/j.mejo.2004.06.022
27. Mo Y, Zhao W, Huang D, Zhao F, Bai M (2009) Nanotribological properties of precision-controlled regular nanotexture on H-passivated Si surface by current-induced local anodic oxidation. Ultramicroscopy 109: 247–252. doi: 10.1016/j.ultramic.2008.10.025
28. Avouris P, Hertel T, Martel R (1997) Atomic force microscope tip-induced local oxidation of silicon: kinetics, mechanism, and nanofabrication. Applied Physics Letters 71: 285. doi: 10.1063/1.119521
29. Hu X, Guo T, Fu X (2003) Nanoscale oxide structures induced by dynamic electric field on Si with AFM. Applied surface science 217: 34–38. doi: 10.1063/1.119521
30. Cabrera N, Mott N (1949) Theory of the oxidation of metals. Reports on Progress in Physics 12: 163. doi: 10.1063/1.119521
31. Teuschler T, Mahr K, Miyazaki S, Hundhausen M, Ley L (1995) Nanometer-scale field-induced oxidation of Si (111): H by a conducting-probe scanning force microscope: Doping dependence and kinetics. Applied Physics Letters 67: 3144. doi: 10.1063/1.119521
32. Sugimura H, Nakagiri N (1995) Fabrication of silicon nanostructures through scanning probe anodization followed by chemical etching. Nanotechnology 6: 29. doi: 10.1063/1.119521
33. Dagata J, Inoue T, Itoh J, Yokoyama H (1998) Understanding scanned probe oxidation of silicon. Applied Physics Letters 73: 271–273. doi: 10.1063/1.121777
34. Marchi F, Bouchiat V, Dallaporta H, Safarov V, Tonneau D, et al. (1998) Growth of silicon oxide on hydrogenated silicon during lithography with an atomic force microscope. Journal of Vacuum Science & Technology B: Microelectronics and Nanometer Structures 16: 2952.

doi: 10.1063/1.121777

35. Stievenard D, Fontaine P, Dubois E (1997) Nanooxidation using a scanning probe microscope: An analytical model based on field induced oxidation. Applied Physics Letters 70: 3272. doi: 10.1063/1.121777
36. Legrand B, Deresmes D, Stievenard D (2002) Silicon nanowires with sub 10 nm lateral dimensions: From atomic force microscope lithography based fabrication to electrical measurements. Journal of Vacuum Science & Technology B: Microelec
37. Avouris P, Hertel T, Martel R (1997) Atomic force microscope tip-induced local oxidation of silicon: kinetics, mechanism, and nanofabrication. Applied Physics Letters 71: 285–287. doi: 10.1063/1.121777
38. Abdullah AM, Sabar DH, Lockman Z (2010) Influence of Room Humidity on the Formation of Nanoscale Silicon Oxide Patterned by Afm Lithography. International Journal of Nanoscience 9: 251–255. doi: 10.1142/s0219581x10006752
39. Hsu HF, Lee CW (2008) Effects of humidity on nano-oxidation of silicon nitride thin film. Ultramicroscopy 108: 1076–1080. doi: 10.1016/j.ultramic.2008.04.025
40. Kuramochi H, Ando K, Yokoyama H (2003) Effect of humidity on nano-oxidation of p-Si (001) surface. Surface science 542: 56–63. doi: 10.1142/s0219581x10006752
41. Jansen H, Gardeniers H, Boer M, Elwenspoek M, Fluitman J (1996) A survey on the reactive ion etching of silicon in microtechnology. Journal of Micromechanics and Microengineering 6: 14. doi: 10.1142/s0219581x10006752
42. Schmuki P, Böhni H, Bardwell J (1995) In situ characterization of anodic silicon oxide films by AC impedance measurements. Journal of the Electrochemical Society 142: 1705. doi: 10.1142/s0219581x10006752
43. Yang P, Lau W, Lai SW, Lo V, Siah S, et al. (2010) The Evolution of Theory on Drain Current Saturation Mechanism of MOSFETs from the Early Days to the Present Day: InTech.
44. Yun M (2000) Investigation of KOH Anisotropic Etching for the Fabrication of Sharp Tips in Silicon-on-Insulator (SOI) Material. Journal-Korean physical society 37: 605–610.
45. Philipsen HGG, Kelly JJ (2009) Influence of chemical additives on the surface reactivity of Si in KOH solution. Electrochimica acta 54: 3526–3531.
46. Zubel I, Barycka I, Kotowska K, Kramkowska M (2001) The effect of organic and inorganic agents on silicon anisotropic etching process.

Sensors and Actuators A: Physical 87: 163–171.

47. Seidel H, Csepregi L, Heuberger A, Baumgartel H (1990) Anisotropic etching of crystalline silicon in alkaline solutions. J Electrochem Soc 137: 3612–3625.
48. Shikida M, Sato K, Tokoro K, Uchikawa D (2000) Differences in anisotropic etching properties of KOH and TMAH solutions. Sensors and Actuators A: Physical 80: 179–188. doi: 10.1016/s0924-4247(99)00264-2
49. Biswas K, Kal S (2006) Etch characteristics of KOH, TMAH and dual doped TMAH for bulk micromachining of silicon. Microelectronics journal 37: 519–525. doi: 10.1016/s0924-4247(99)00264-2
50. Youn S, Kang C (2006) Effect of nanoscratch conditions on both deformation behavior and wet-etching characteristics of silicon (1 0 0) surface. Wear 261: 328–337. doi: 10.1016/s0924-4247(99)00264-2
51. Pennelli G, Piotto M, Barillaro G (2006) Silicon single-electron transistor fabricated by anisotropic etch and oxidation. Microelectronic engineering 83: 1710–1713. doi: 10.1016/s0924-4247(99)00264-2
52. Zubel I, Kramkowska M (2001) The effect of isopropyl alcohol on etching rate and roughness of (1 0 0) Si surface etched in KOH and TMAH solutions. Sensors and Actuators A: Physical 93: 138–147. doi: 10.1016/s0924-4247(99)00264-2
53. Zubel I, Kramkowska M (2002) The effect of alcohol additives on etching characteristics in KOH solutions. Sensors and Actuators A: Physical 101: 255–261.
54. Merlos A, Acero M, Bao M, Bausells J, Esteve J (1993) TMAH/IPA anisotropic etching characteristics. Sensors and Actuators A: Physical 37: 737–743.
55. Yang CR, Chen PY, Yang CH, Chiou YC, Lee RT (2005) Effects of various ion-typed surfactants on silicon anisotropic etching properties in KOH and TMAH solutions. Sensors and Actuators A: Physical 119: 271–281.
56. Camon H, Moktadir Z (1997) Simulation of silicon etching with KOH. Microelectronics journal 28: 509–517.
57. Campbell S, Cooper K, Dixon L, Earwaker R, Port S, et al. (1999) Inhibition of pyramid formation in the etching of Si p (100) in aqueous potassium hydroxide-isopropanol. Journal of Micromechanics and Microengineering 5: 209.
58. Dehzangi A, Larki F, Hassan J, Hutagalung SD, Saion EB, et al. (2013) Fabrication of p-Type Double Gate and Single Gate Junctionless Silicon Nanowire Transistor by Atomic Force Microscopy Nanolithography.

Nano Hybrids 3: 93–113. doi: 10.4028/www.scientific.net/nh.3.93

59. Haiss W, Raisch P, Bitsch L, Nichols RJ, Xia X, et al. (2006) Surface termination and hydrogen bubble adhesion on Si (1 0 0) surfaces during anisotropic dissolution in aqueous KOH. Journal of Electroanalytical Chemistry 597: 1–12. doi: 10.4028/www.scientific.net/nh.3.93
60. Biswas K, Das S, Kal S (2006) Analysis and prevention of convex corner undercutting in bulk micromachined silicon microstructures. Microelectronics journal 37: 765–769. doi: 10.4028/www.scientific.net/nh.3.93
61. Martinez R, Losilla N, Martinez J, Huttel Y, Garcia R (2007) Patterning polymeric structures with 2 nm resolution at 3 nm half pitch in ambient conditions. Nano letters 7: 1846–1850. doi: 10.1021/nl070328r
62. Tello M, Garcia R, Martín-Gago JA, Martínez NF, Martín-González MS, et al. (2005) Bottom–Up Fabrication of Carbon-Rich Silicon Carbide Nanowires by Manipulation of Nanometer-Sized Ethanol Menisci. Advanced Materials 17: 1480–1483. doi: 10.4028/www.scientific.net/nh.3.93
63. Gordon A, Fayfield R, Litfin D, Higman T (1995) Mechanisms of surface anodization produced by scanning probe microscopes. Journal of Vacuum Science & Technology B: Microelectronics and Nanometer Structures 13: 2805–2808. doi: 10.1116/1.588270
64. Dagata J, Perez-Murano F, Martin C, Kuramochi H, Yokoyama H (2004) Current, charge, and capacitance during scanning probe oxidation of silicon. I. Maximum charge density and lateral diffusion. Journal of applied physics 96: 2386. doi: 10.1116/1.588270

# Chapter 2

# BINARY DNA NANOSTRUCTURES FOR DATA ENCRYPTION

Ken Halvorsen[1, 2], Wesley P. Wong[1, 2*]

[1]Immune Disease Institute/Program in Cellular and Molecular Medicine, Harvard Medical School and Children's Hospital Boston, Boston, Massachusetts, United States of America

[2]Department of Biological Chemistry and Molecular Pharmacology, Harvard Medical School, Boston, Massachusetts, United States of America

## ABSTRACT

We present a simple and secure system for encrypting and decrypting information using DNA self-assembly. Binary data is encoded in the geometry of DNA nanostructures with two distinct conformations. Removing or leaving out a single component reduces these structures to an encrypted solution of ssDNA, whereas adding back this missing "decryption key" causes the spontaneous formation of the message through self-assembly, enabling rapid read out via gel electrophoresis. Applications include authentication, secure messaging, and barcoding.

## INTRODUCTION

As the blueprint for all living things, DNA has a remarkable ability to robustly store and relay information. Two salient features of DNA make this possible: its modular construction from four distinct bases (A, C, G, T) whose sequence determines the genetic code, and the specific base pairing between complementary bases that enables hybridization into a double-stranded complex. These features of DNA have been exploited to perform computations and process information [1]–[6], including the hiding and encryption of secret messages [7]–[10]. Not only has DNA formed a foundation for the field of biomolecular computing, but the robustness of DNA-base pairing has also led to its use as a *programmable* structural material to construct objects with nanoscale features [11]. Recently, the accessibility and versatility of DNA nanotechnology has been remarkably increased with an approach referred to as DNA origami[12]. With a carefully designed collection of oligonucleotides, DNA can self-assemble into 2D and 3D shapes of stunning complexity [12]–[14], some of which incorporate sensing and actuation [15], [16]. Combining the concepts of DNA as an information processing molecule and as a structural material, we have developed a simple and powerful approach for encoding and encrypting information.

In previous work, we used DNA origami methods to construct a nanoscale mechanical switch[17], which we used to study the force-dependence of molecular interactions at the single-molecule level. This switch could be in one of two states, looped or unlooped, and we observed that these conformational differences were clearly resolvable in an agarose gel (Figure 1a). Thus, while other schemes for representing binary data using DNA have been presented [8],[18], [19], here we focus on using the geometric conformation of DNA nanostructures to encode binary values, due to the ease of both encoding and decoding information using this approach. We have made three distinct realizations of this concept, as demonstrated in Figure 1. These nanoscale structures can switch between two distinct states, acting as a "mechanical bit" to enable the storage and processing of information (analogous to mechanical relays in the earliest digital computers). These mechanical bits can be prepared in either state (0 or 1), and transitions between these states could be controlled using chemical or physical means (e.g. by changing the presence or absence of a critical molecular component, by varying the temperature, by interactions with light [20], or by applying mechanical force [17], [21]). Importantly, the state of these bits can be read out in minutes using gel electrophoresis, or faster using single-molecule imaging and manipulation techniques, and multiple bits can be represented with different lengths of DNA to facilitate multiplexed information processing and readout.

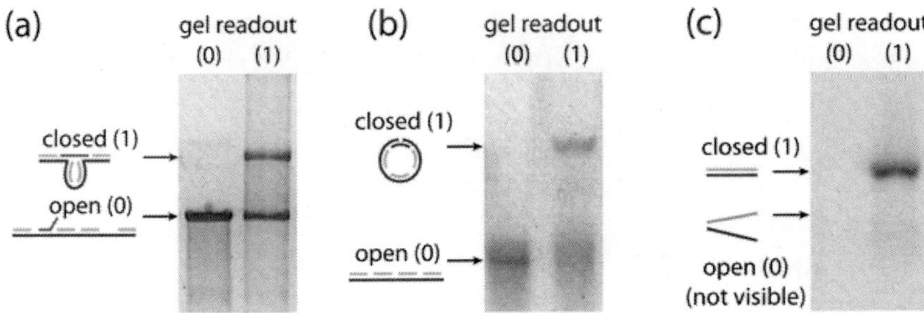


Figure 1. DNA as a binary switch.

The conformations of two-state DNA nanostructures can represent bits through open or closed states, representing 0 and 1 respectively. We demonstrate this concept with three different implementations: a) a self-assembled construct with an addressable loop closure as described previously [17], b) a switchable circular/linear construct, and c) a double-stranded/single-stranded segment.

doi:10.1371/journal.pone.0044212.g001

Each mechanical bit is formed via DNA self-assembly, and can be encrypted by omitting a critical component of its structure (e.g. a key single-stranded DNA molecule) that reduces it to an unstructured mixture of oligonucleotides. Messages encrypted as a collection of such bits are difficult to decipher since the 0 bits and 1 bits are nearly indistinguishable mixtures of oligonucleotides that are identical in all but sequence. On the other hand, decryption with a key is very easy–simply adding the missing component triggers the self-assembly of these nanoscale mechanical bits into their unencrypted forms. The separation of each mechanical bit into two parts or two "keys" forms an asymmetric encryption system. This system has the "public key" property if the key is distributed physically, since one key cannot be readily determined from the other without knowledge of the sequence. Furthermore, suitable countermeasures, such as adding "distractor" oligos to the physical encryption key to obscure information, can make the decryption sequence difficult to obtain.

As an example, let us consider how Alice could send an encrypted message to Bob (Figure 2) using the linear binary switch shown in Figure 1C. Suppose she would like to send a three bit message, such as "101". First, Bob must generate the appropriate DNA encryption and decryption keys for each bit. To distinguish between bits, he chooses each to be a different DNA length, (e.g 20 bases, 30 bases and 40 bases), and then generates 3 equal length oligos for each bit (A, A', and B), two of which are complementary and hybridize together (A and A') and one of which is inert (B) (see Materials and Methods for details on oligo design). Then Bob makes vials of A and B for each bit available, which represent the 1 and 0 values, respectively, while keeping the vials of oligo A' private. Together, the vials of A and B oligos form the*encryption key*, which can be used by anyone to encrypt a message. To send a message to Bob, Alice would mix either A (for a 1) or B (for a 0) for each bit into a single vial and send this mixture to Bob over a public channel. At this point, only Bob can decrypt the message by mixing in the private *decryption key* (the set of A' oligos) and running a gel–even Alice has no way to decrypt her own message once it's been made.

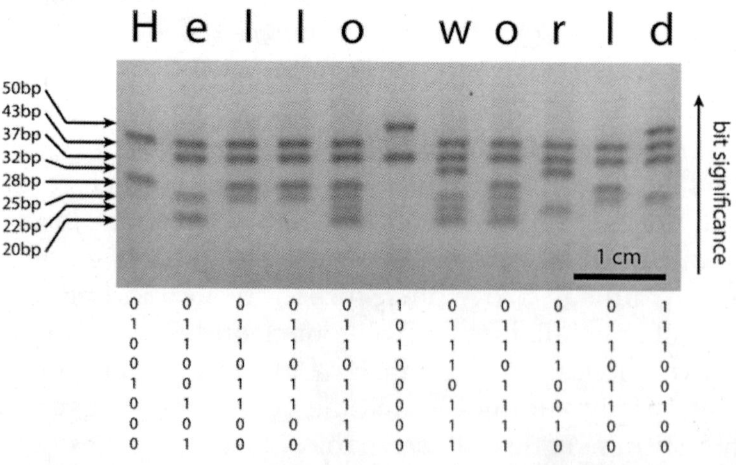


Figure 2. Conceptualization of DNA encryption and decryption.

Alice prepares her message by mixing together oligos that correspond to either a binary 0 or 1 for each bit. This mixture is sent to Bob through a public channel, who decrypts the message by adding the DNA decryption key. This causes the message to self-assemble, enabling rapid read out by gel electrophoresis.

doi:10.1371/journal.pone.0044212.g002

## Results and Discussion

We experimentally demonstrated this encryption scheme for an 8-bit encoding by using 8 different lengths of DNA strands to represent the individual bits. We encoded a plaintext message using an 8-bit binary ASCII encoding with the 8th bit as an even parity bit, and encrypted the message into an ordered sequence of DNA mixtures, one for each letter. These mixtures were then decrypted by mixing them with the private key oligos and read out by immediately running an agarose gel. Reading each gel lane from top to bottom, the decrypted message "Hello world" becomes unambiguously clear (Figure 3).

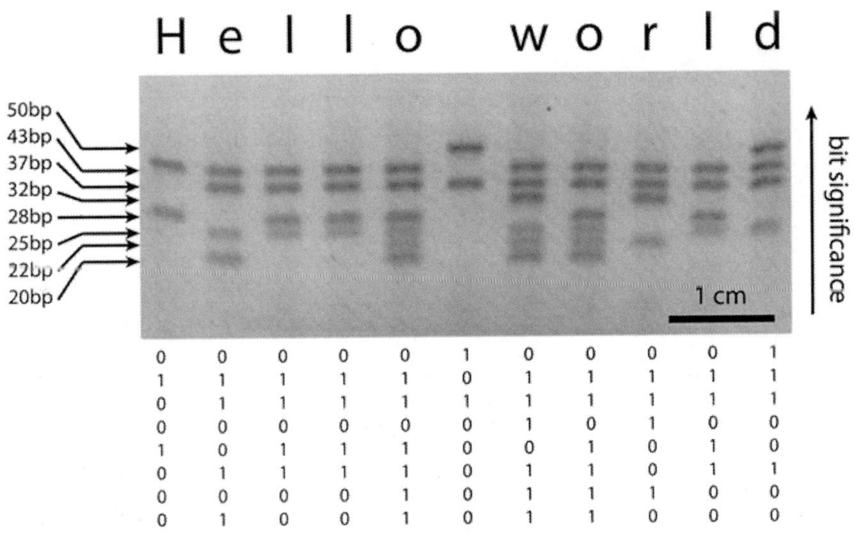


Figure 3. Decoding a binary message on a gel.

Each lane of the gel contains a mixture of oligonucleotides which together form an 11 byte binary ASCII message which reads "Hello world". The bit strings are read from top to bottom with the most significant bit being the largest DNA segment. Presence of a band indicates a binary 1, while absence of a band indicates a binary 0. All lanes have the same amount of DNA present, but only the double stranded pieces dye.

doi:10.1371/journal.pone.0044212.g003

One of the most intriguing aspects of encrypting messages with DNA as described is the difficulty for an interceptor or attacker to decrypt it. At a minimum, decryption attempts require possession of the physical message as well as technical skill, laboratory equipment, and time. Since the message is encrypted physically, the transmission of data can be well controlled, and even copying the encrypted message poses a significant technical challenge. Unlike encryption schemes that rely on mathematical algorithms, our biochemical based encryption is not directly vulnerable to increasing computational power. With physical decryption, the number of decryption attempts is limited by the availability of physical material comprising the message, which is depleted with each attempt. In fact, the message could theoretically be reduced to enable only a single decryption attempt.

Attempts to crack the message without the decryption key would be difficult, especially for an attacker limited to the same resources as the intended recipient (i.e. electrophoresis and mixing equipment). For example, a näive brute force attack to find the decryption key would require the physical generation and testing of an astronomical number of possible keys, with $4^N$distinct possibilities, or $10^{155}$ for our simple 8-bit encoding scheme (and an approximately 1 in $10^{63}$ chance of guessing a decryption key if we allow for 25% mismatched bases [8], [22]). However, aside from the correct decryption key, there could also exist a pseudo key or set of pseudo keys that could allow an attacker to distinguish between individual mixtures, providing a toehold for unauthorized decryption using language statistics on a large set of messages. Additionally, if the attacker has access to more sophisticated approaches, including sequencing techniques, DNA profiling with microarrays, or the ability to make libraries of oligos to test multiple keys at a time, then the security will be reduced. Fortunately, many of these approaches can be impeded by implementing the appropriate countermeasures, as described below. In addition, it is interesting to note that many of the potential ways to crack the code may involve significant technical expertise, expensive laboratory equipment, or time consuming processes, giving rise to a large asymmetry in effort compared to encrypting and decrypting the message with the proper keys.

To maximize encryption security, a variety of countermeasures can be used that comprise a two pronged approach: 1) limiting access to physical information about the message (e.g. the oligo sequences), and 2) making this physical information difficult to decipher. One simple countermeasure is to limit the physical amount of material comprising the message, as mentioned above. Additionally, we can impede chemical analysis by modifying the ends of the oligos to prevent the chemical conjugation required for some sequencing and profiling techniques. One important countermeasure against unauthorized decryption that falls into both categories is the addition of noise by mixing "distractor strands" into the physical message (described in detail elsewhere [8], [23]). Adding oligonucleotides with a similar length and composition as the message strands but with different sequences increases the time and effort required to obtain the sequences of the oligonucleotides within the encrypted mixture, which is already a challenge due to their short lengths [24]–[26]. Also, distractor strands make deciphering the message more difficult since the true message can be obscured by noise designed to like the signal, or vice versa [23]. For a public key system, special care should be taken in designing the distractor strands to ensure that the two values for each bit are difficult to distinguish using a pseudo key or other biochemical techniques (e.g. mass spectrometry). For a private key system, distractor strands can be very effective at deterring statistical analysis, as each occurrence of a given letter could be mixed with a different set of distractor strands. In fact, false messages could be encoded into the mixture with distractor strands that represent "false encryption keys", and the attacker would have little way to identify the true message without knowing the specific decryption key sequence. Alternatively, to make the signal look more like the noise, DNA structures could be pre-encrypted with standard computer encryption algorithms. Furthermore, cracking the code becomes increasingly challenging the more distractor strands are added, and the more bits per message are encoded.

While our demonstration showed 1 byte (8 bits) of data storage per mixture (and per gel lane), the amount of data stored and read out could be readily increased. If DNA lengths were optimized to be evenly spaced on the gel, a resolution of 1 mm would give 10 bits/cm of gel length, corresponding to roughly 8 bytes of data for a single short gel lane, 10–20 bytes for longer gels, and up to 125 bytes (~1000 bits) for more elaborate sequencing gels with base pair resolution [27]. Expanding to a few bytes could enable the transmission of entire words or short messages in a single mixture, especially if a more efficient character encoding scheme were used (e.g. the 5 bit Baudot code), or a word-based encoding scheme (e.g. 10 bytes could encode eight words of a 1000 word vocabulary). In addition, data density could be dramatically increased by combining multiple messages within a single mixture, with each message associated with a different decryption key. We note that these approaches will improve security by making statistical analysis of the message difficult.

In summary, we have developed a novel technique for encoding binary information in two-state DNA nanostructures, for encrypting and decrypting this information using self-assembly, and for rapidly reading this information back out with gel electrophoresis. As the state of such nanostructures can be used to report molecular events (such as the rupture of intermolecular bonds [17]), additional applications of this work beyond information security include the characterization of molecular interactions with a multiplexed gel readout. More directly, this approach provides a relatively simple and inexpensive way to send secure, and potentially hidden, messages over a public channel. Unlike similar DNA encryption methods that require specialized laboratory work (e.g. PCR, sequencing, cloning) that can take hours to days[7]–[10], our technique enables encrypting and decrypting messages in just minutes, requiring as little as disposable droppers and a no-prep bufferless (even handheld) gel system (e.g. Invitrogen E-gel systems). This convenience and simplicity also opens the technology to other applications such as authentication and barcoding. For example, increasing storage capacity to 40 bits would enable storage of the ubiquitous 12 digit Universal Product Code (UPC). Several methods have been demonstrated for storing printed DNA on paper [28],

[29], which combined with our quick readout method could enable a new level of security in product identification. In fact, tagging with DNA inks has recently found commercial use (DNA Technologies, Halifax, Canada) in anti-counterfeiting efforts for a variety of products including sports memorabilia, artwork, pharmaceuticals, and luxury goods. As previously proposed, DNA barcodes may also find use in labeling of liquids such as paint and oil, or possibly even food [8]. As another example, it could provide a simple and inexpensive way for pharmaceutical companies to discretely label drugs with production or expiration information, even at the level of edible encrypted barcodes on individual tablets. This is particularly timely in light of recently introduced government mandates (e.g. California E-Pedigree Law) to serialize pharmaceuticals and to ultimately "track and trace" them, since our method offers a way to identify and authenticate drugs, reducing theft and counterfeiting. One could also envision personal identification cards (e.g. driver's license, passports) being printed with DNA markers on them as an additional prevention against fraud or identity theft, or even using one's own genomic DNA as an authentication key.

## Materials and Methods

We designed and purchased oligos (Bioneer, Inc.) to represent 8 different bits, which would be about evenly spaced on a 4% agarose gel. The lengths we chose were 20 nt, 22 nt, 25 nt, 28 nt, 32 nt, 37 nt, 43 nt, and 50 nt. For each length, 3 oligos were purchased: a randomly generated sequence, it's complementary strand, and a random set with arbitrary bases. We denote these as set A, A', and B respectively.

To encode messages, we first converted our plain text message "Hello world" into binary code using 8 bit ASCII character encoding with the 8th bit as an even parity bit for error checking. Since we have 8 bits total, each letter was prepared using a mixture of A and B oligos to represent the 0 s and 1 s. As an example, the "H" in "Hello world" has an 8 bit binary representation of 01001000. The least significant bit is encoded in the smallest (20 nt) oligo, which we will denote oligo 1. To encode the "H", we mix oligos 4 and 7 from set A with oligos 1, 2, 3, 5, 6, and 8 from set B.

For decoding, the encoded message mixtures were individually mixed with the entire set A' of oligos in a buffer solution (1× Buffer 4,

New England Biolabs) and loaded into a gel (within minutes). We ran an automated precast 4% agarose gel (E-gel, Invitrogen) containing a proprietary dye (with characteristics remarkably similar to Sybr gold) for 15 minutes and took a picture immediately after. The binary representation of each gel lane could be read directly from top to bottom.

## Acknowledgments

The authors would like to thank Daniel Cheng, Ted Feldman, Darren Yang, and Mounir Koussa for helpful discussions.

## Author Contributions

Conceived and designed the experiments: KH WW. Performed the experiments: KH. Analyzed the data: KH WW. Wrote the paper: KH WW.

## REFERENCES

1. Adleman L (1994) Molecular computation of solutions to combinatorial problems. Science 266: 1021–1024. doi: 10.1126/science.7973651
2. Lipton R (1995) DNA solution of hard computational problems. Science 268: 542–545. doi: 10.1126/science.7725098
3. Guarnieri F, Fliss M, Bancroft C (1996) Making DNA add. Science 273: 220–223. doi: 10.1126/science.273.5272.220
4. Ouyang Q, Kaplan P, Liu S, Libchaber A (1997) DNA solution of the maximal clique problem. Science 278: 446–449. doi: 10.1126/science.278.5337.446
5. Braich R, Chelyapov N, Johnson C, Rothemund P, Adleman L (2002) Solution of a 20-variable 3-sat problem on a DNA computer. Science 296: 499–502. doi: 10.1126/science.1069528
6. Qian L, Winfree E (2011) Scaling up digital circuit computation with DNA strand displacement cascades. Science 332: 1196. doi: 10.1126/science.1200520
7. Clelland C, Risca V, Bancroft C (1999) Hiding messages in DNA microdots. Nature 399: 533–534. doi: 10.1038/21092
8. Leier A, Richter C, Banzhaf W, Rauhe H (2000) Cryptography with DNA binary strands. Biosystems 57: 13–22. doi: 10.1016/S0303-2647(00)00083-6
9. Tanaka K, Okamoto A, Saito I (2005) Public-key system using DNA as a one-way function for key distribution. Biosystems 81: 25–29. doi: 10.1016/j.biosystems.2005.01.004

10. Cui G, Qin L, Wang Y, Zhang X (2008) An encryption scheme using DNA technology. Bio-Inspired Computing: Theories and Applications 2008 37–42. doi: 10.1109/bicta.2008.4656701
11. Seeman N (2007) An overview of structural DNA nanotechnology. Molecular biotechnology 37: 246–257. doi: 10.1007/s12033-007-0059-4
12. Rothemund P (2006) Folding DNA to create nanoscale shapes and patterns. Nature 440: 297–302. doi: 10.1038/nature04586
13. Yin P, Hariadi R, Sahu S, Choi H, Park S, et al. (2008) Programming DNA tube circumferences. Science 321: 824–826. doi: 10.1126/science.1157312
14. Douglas S, Dietz H, Liedl T, Högberg B, Graf F, et al. (2009) Self-assembly of DNA into nanoscale three-dimensional shapes. Nature 459: 414–418. doi: 10.1038/nature08016
15. Andersen E, Dong M, Nielsen M, Jahn K, Subramani R, et al. (2009) Self-assembly of a nanoscale DNA box with a controllable lid. Nature 459: 73–76. doi: 10.1038/nature07971
16. Douglas S, Bachelet I, Church G (2012) A logic-gated nanorobot for targeted transport of molecular payloads. Science 335: 831–834. doi: 10.1126/science.1214081
17. Halvorsen K, Schaak D, Wong W (2011) Nanoengineering a single-molecule mechanical switch using DNA self-assembly. Nanotechnology 22: 494005. doi: 10.1088/0957-4484/22/49/494005
18. Roweis S, Winfree E, Burgoyne R, Chelyapov N, Goodman M, et al. (1998) A sticker-based model for DNA computation. Journal of Computational Biology 5: 615–629. doi: 10.1089/cmb.1998.5.615
19. Yan H, LaBean T, Feng L, Reif J (2003) Directed nucleation assembly of DNA tile complexes for barcode-patterned lattices. Proceedings of the National Academy of Sciences 100: 8103. doi: 10.1073/pnas.1032954100
20. Schäfer C, Eckel R, Ros R, Mattay J, Anselmetti D (2007) Photochemical single-molecule affinity switch. Journal of the American Chemical Society 129: 1488–1489. doi: 10.1021/ja067734h
21. Quek S, Kamenetska M, Steigerwald M, Choi H, Louie S, et al. (2009) Mechanically controlled binary conductance switching of a single-molecule junction. Nature Nanotechnology 4: 230–234. doi: 10.1038/nnano.2009.10
22. Lee I, Dombkowski A, Athey B (2004) Guidelines for incorporating non-perfectly matched oligonucleotides into target-specific hybridization probes for a DNA microarray. Nucleic acids research 32: 681–690. doi: 10.1093/nar/gkh196
23. Gehani A, LaBean T, Reif J (2004) DNA-based cryptography. Aspects of Molecular Computing : 34–50.
24. Oberacher H, Mayr B, Huber C (2004) Automated de novo sequencing of nucleic acids by liquid chromatography-tandem mass spectrometry. Journal of the American Society for Mass Spectrometry 15: 32–42. doi: 10.1016/j.jasms.2003.09.005

25. Oberacher H, Pitterl F (2009) On the use of esi-qqtof-ms/ms for the comparative sequencing of nucleic acids. Biopolymers 91: 401–409. doi: 10.1002/bip.21156

26. Farand J, Gosselin F (2009) De novo sequence determination of modified oligonucleotides. Analytical chemistry 81: 3723–3730. doi: 10.1021/ac802452p

27. França L, Carrilho E, Kist T (2002) A review of DNA sequencing techniques. Quarterly reviews of biophysics 35: 169–200. doi: 10.1017/S0033583502003797

28. Kawai J, Hayashizaki Y (2003) DNA book. Genome research 13: 1488–1495. doi: 10.1101/gr.914203

29. Hashiyada M (2004) Development of biometric DNA ink for authentication security. The Tohoku Journal of Experimental Medicine 204: 109–117. doi: 10.1620/tjem.204.109

# Chapter 3

# NANOSTRUCTURED 3D CONSTRUCTS BASED ON CHITOSAN AND CHONDROITIN SULPHATE MULTILAYERS FOR CARTILAGE TISSUE ENGINEERING

Joana M. Silva[1,2], Nicole Georgi[3,4], Rui Costa[1,2], Praveen Sher[1,2], Rui L. Reis[1,2], Clemens A. Van Blitterswijk[3], Marcel Karperien[3,4], Joa˜o F. Mano[1,2*]

[1]3B's Research Group – Biomaterials, Biodegradables and Biomimetics, University of Minho, Headquarters of the European Institute of Excellence on Tissue Engineering and Regenerative Medicine, Taipas, Guimara˜es,Portugal

[2]ICVS/3B's – PT Government Associate Laboratory, Braga/Guimara˜es, Portugal, [3]Department of Tissue Regeneration, MIRA – Institute for Biomedical Technology and Technical Medicine, University of Twente, Enschede, The Netherlands

[4]Department of Developmental BioEngineering, MIRA – Institute for Biomedical Technology and Technical Medicine, University of Twente, Enschede, The Netherlands

## ABSTRACT

Nanostructured three-dimensional constructs combining layer-by-layer technology (LbL) and template leaching were processed and evaluated as possible support structures for cartilage tissue engineering. Multilayered constructs were formed by depositing the polyelectrolytes chitosan (CHT) and chondroitin sulphate (CS) on either bidimensional glass surfaces or 3D packet of paraffin spheres. 2D CHT/CS multi-layered constructs proved to support the attachment and proliferation of bovine chondrocytes (BCH). The technology was transposed to 3D level and CHT/CS multi-layered hierarchical scaffolds were retrieved after paraffin leaching. The obtained nanostructured 3D constructs had a high porosity and water uptake capacity of about 300%. Dynamical mechanical analysis (DMA) showed the viscoelastic nature of the scaffolds. Cellular tests were performed with the culture of BCH and multipotent bone marrow derived stromal cells (hMSCs) up to 21 days in chondrogenic differentiation media. Together with scanning electronic microscopy analysis, viability tests and DNA quantification, our results clearly showed that cells attached, proliferated and were metabolically active over the entire scaffold. Cartilaginous extracellular matrix (ECM) formation was further assessed and results showed that GAG secretion occurred indicating the maintenance of the chondrogenic phenotype and the chondrogenic differentiation of hMSCs.

## INTRODUCTION

Articular cartilage is an avascular, alymphatic, aneural, anisotropic tissue with limited capacity to regenerate [1], [2]. Due to these articular cartilage properties, tissue engineering approaches are needed to treat millions of people which suffer from traumatic injuries and degenerative cartilage diseases. A wide range of clinical options emerged to repair these lesions such as micro-fracture, micro-drilling auto and allografts, among others [3]. However, these treatments present some limitations, e.g. availability of sufficient cells for repair, quality and quantity of repaired tissue, and thereby fail to produce long-lasting repair [4], [5].

Tissue Engineering (TE) has appeared as a new method, which offers advantages when compared with current treatments [5], [6]. Scaffolds play an important role in TE strategies because they provide the initial support structure, guiding the differentiation and development of the cartilaginous tissue [7]–[9]. Typically native tissues exhibit a hierarchical organization from the nano- to the macro-scale levels which is difficult to achieve in conventional scaffolds. Thus, the control from the nano-sizes to macroscale of scaffold is of great interest because offers the possibility of developing structures with further capabilities. These capabilities include the fabrication of hierarchical-organized structures, the control of cell behaviour at the nano-level and the inclusion of other functionalities, such as the possibility of incorporate bioactive molecules, or tune the mechanical and degradation behaviour of the scaffold. This structures can be achieved by layer-by-layer (LbL) methodology, a versatile technique that permits to fabricate nanostructured multilayered films using a variety of polyelectrolytes [10]–[12]. The principle of this technique is based on alternate deposition of polyelectrolytes that will self-organize on the material surface [10]–[13]. The main application of LbL is the build-up of polyelectrolytes multilayers (PEMs) onto flat surfaces [10]–[12]. Just a few works reported the use of LbL to fabricate scaffolds. Such technique may be used to coat free-packet leachable spherical templates [14] or to agglomerate beads [15], leading in both cases to porous structures. In this work we propose the use of an LbL based bottom-up approach to produce three-dimensional (3D) highly porous scaffolds with a nanostructured organization reminiscent of the native extracellular matrix components of cartilage.

Cartilage specific ECM components play an important role in chondrogenesis as well as supporting the chondrogenic phenotype. Among the wide range of materials that has been explored for cartilage TE approaches appears chitosan (CHT) and chondroitin sulphate [6], [16]. CHT, a naturally derived is an excellent candidate for polycation due to its structural characteristics similar to glycosaminoglycan's (GAGs) and ability to support chondrogenic activity as well as Cartilage ECM expression by chondrocytes [17], [18]. Chondroitin sulphate (CS) has a high negative charge density and it is the major GAG component of native cartilage tissue and it is reported its benefits for osteoarthritis as well as its ability to increase the production of ECM matrix a and capacity to induce the differentiation of multipotent stromal cells [1], [19]–[21]. This combination was already used in LbL methodology, however from our knowledge we reported from the first time the use of these polyelectrolytes for a 3D porous construct only based in PEMs for cartilage TE approaches [22].

The aim of this work is to prepare nanostructured 3D constructs, based on the LbL methodology, studying its effect on cartilage TE. For the proof of concept the build-up of CHT/CS PEMs onto flat surfaces was firstly characterized using quartz crystal microbalance (QCM). The biological performance was evaluated with a cell culture of primary bovine chondrocytes (BCH). The biological performance of highly porous nanostructured 3D scaffolds was also evaluated using BCH and multipotent bone marrow derived stromal cells (hMSCs). The maintenance of chondrogenic phenotype and the differentiation of hMSCs were also investigated.

## MATERIALS AND METHODS

### Materials

Chitosan (CHT) of medium molecular weight ($M_w$ 190–310 kDa, 75–85% degree of deacetylation, viscosity 200–800 cP) and chondroitin-4-sulphate (CS) ($M_w$ 50–100 kDa) were purchased from Sigma Aldrich. Chitosan was purified by recrystallization. Paraffin wax spheres with Ø 200 µm were purchased from Jojoba Desert Whale (Tucson, USA) and then modified with polyethylene imine (PEI) (Sigma-Aldrich, Mw 750 000). Glass coverslips with Ø13 mm (L4097-3) were purchased from Agar Scientific. Lysozyme from chicken egg white (lyophilized powder ≈10000 U/mg stored at 4°C) and hyaluronidase Type VIII (300 U/mg stored at −20°C) were purchased from Sigma-Aldrich.

## METHODS

### CHT/CS film build-up

The build-up process of CHT/CS PEMs was followed in situ by quartz crystal microbalance with dissipation monitoring (QCM-Dissipation, Q-Sense, Sweden), using a gold coated sensor excited at a fundamental frequency of 5 MHz and at seventh overtone (35 MHz). The crystals were cleaned in an ultrasound bath at 30°C using successive acetone, ethanol and isopropanol. Adsorption took place with a constant flow rate of 50 μL $min^{-1}$.

The CHT (0.15% (w/v) in 1% acetic acid/ 0.15 M NaCl, pH = 5.5) solution was pumped into the system for 10 min to allow the adsorption equilibrium at the crystal surface. After rinsing with 0.15 M NaCl (10 min), the same procedure was followed for the deposition of CS (0.15% (w/v) in 1% acetic acid/ 0.15 M NaCl, pH = 5.5). The steps were repeated to the desire number of layers. The frequency and dissipation were monitored in real time. The thickness of the film was estimated using the Voigt model through the Q-Tools Software, from Q-Sense [23].

## LbL assembly in 2D surfaces

The CHT/CS PEMs were deposited onto glass coverslips. The glass coverslips were placed in 70% (v/v) ethanol for 2 hours and then immersed in 0.15 M NaCl for 10 min. After these two steps the glass coverslips were dried using nitrogen flow. The multilayered film build-up started by immersing first the substrate in CHT during 10 min followed by the immersion in 0.15 M NaCl solution during 5 min. Then the coverslips were dipped in CS solution for 10 min, followed by immersion in 0.15 M NaCl over 5 min. These four steps allowed the assembling of one double layer. The process was repeated until 10 double layers were achieved.

## Scaffolds production by LbL

The PEMs were constructed onto free-packet paraffin spheres previously modified with PEI. Paraffin spheres modified with PEI were chosen as the porogen and 150 mg of them placed into a modified cylindrical container, with a porous base. Drop wise addition of polyelectrolyte solutions and washing solutions over the top of assembly was done to form 10 double layers. The coated structure was placed in dichloromethane (DCM) to leach out the paraffin. After the leaching the samples were freeze dried.

## MORPHOLOGY

The morphology of the scaffolds after the leaching process and immersion in DCM was assessed by optical microscopy, using the Axioplan Imager Z1 microscope (Zeiss). Freeze-dried scaffolds were also observed by scanning electronic microscopy (SEM), using a Philips XL 30 ESEM-FEG operated at 15 kV accelerating voltage. Surface morphology of the coated glass coverslips was also observed using the same equipment at 7.5 kV accelerating voltage. All the samples were sputtered with a conductive gold layer, using a sputter coater (Cressington) for 40 s at a current of 40 mA.

### Fourier transform infrared (FTIR) spectroscopy

FTIR measurements were recorded using an IRPrestige-21 spectrophotometer, by averaging 34 individual scans over the range 4400 $cm^{-1}$ to 400 $cm^{-1}$. The samples were prepared in potassium bromide (KBr) discs.

### Swelling test

The water uptake ability of the scaffolds with known weight was determined by soaking them in phosphate buffered saline solution (PBS, Gibco) at pH = 7.4 up to 3 days at 37°C. The swollen scaffolds were removed at predetermined time points (t = 15 min, 30 min, 1 h, 2 h, 3 h, 4 h, 5 h, 1 day, 2 days and 3 days). After removing the excess water using a filter paper (Whatman Pergamyn Paper), the scaffolds were weighed with an analytical balance (Scaltec, Germany). The water uptake was calculated, where $W_w$ and $W_d$ are the weights of swollen and dried scaffold, respectively.

$$Wateruptake\% = \frac{Ww - Wd}{Wd} \times 100 \quad (1)$$

### Enzymatic Degradation

The enzymatic degradation test was performed to evaluate the degradation profile of the scaffolds in simulated physiological environments. Scaffolds were placed at 37°C in PBS solution (pH = 7.4) or in enzymatic solution containing 2 $mg.ml^{-1}$ of lysozyme and 0.33

mg.ml$^{-1}$ of hyaluronidase (pH 7.4) (19). PBS and enzymatic solution were changed every third day (19). At predetermined time intervals, t = 3, 7 and 14 days the scaffolds were washed with distilled water to remove the salts. Then the scaffolds were immersed in ethanol 100% and dried for 1 day at room temperature. The percentage of weight loss ($W_L$) was calculated, where $W_i$ and $W_f$ are the weights of dry scaffold and after incubation in PBS or enzymatic solution, respectively. $W_L = \frac{Wi - Wf}{Wi} \times 100$ (2)

## Mechanical Test

Compression tests were carried out using dynamic mechanical analysis (DMA), using Tritec 2000B equipment (Triton Technology, UK) to characterize the mechanical properties of cylindrical scaffolds in both the dry and wet states. The sizes of the samples were measured using a digital micrometre. Prior to any measurements in the wet state the scaffolds were immersed in PBS until equilibrium was reached. The measurement was carried out at 37°C under full immersion of the sample in liquid bath (PBS) placed in a Teflon® reservoir. Experiments were carried out in compression mode following cycles of increasing frequency ranging from 0.1 to 15 Hz, with constant strain amplitude of 30 µm. The frequency range chosen covers the characteristic timescales of the periodic loads felt by the scaffold *in vivo*(e.g. typical frequency of skeletal movement). The high frequency limit used in this study should provide information about the viscoelastic properties for the equivalent of short times[24].

## Bovine articular chondrocytes and human mesenchymal stem cells culture

BCH cells were harvested from a patellar-femoral groove of calf legs and isolated by 0.2% collagenase overnight digestion (37°C) [25]. hMSCs were selected by adherence from the bone marrow of human donors undergoing total hip replacement [26]. Ethical approval has been obtained from a local medical ethical committee. The isolated BCH were washed, centrifuged and re-suspended in chondrocyte proliferation medium containing dulbecco's modified eagle medium (DMEM, Invitrogen, USA), fetal bovine serum (FBS, 10%, Sigma-

Aldrich), non-essential aminoacids (0.1 mM, Sigma-Aldrich), penicillin/streptomycin (100 U/100 μg.mL$^{-1}$, Invitrogen), proline (0.4 mM, Sigma-Aldrich) and Ascorbic acid 2-phosphate (ASAC, 0.2 mM, Invitrogen) in a humidified atmosphere with 5% $CO_2$ and at 37°C. hMSCs were also washed, centrifuged and re-suspended in MSCs proliferation medium containing alpha modified eagle›s medium (α-MEM, Invitrogen, USA), fetal bovine serum (FBS, 10% Sigma-Aldrich), penicillin/streptomycin (100 U/100 μg.mL$^{-1}$, Invitrogen), Glutamine (2 mM, Sigma-Aldrich), basic fibroblast growth factor (bFGF, 1 ng.mL−1, Sigma Aldrich) and ASAC (0.2 mM, Invitrogen) in a humidified atmosphere with 5% CO2 and at 37°C.

BCH and hMSCs were seeded in tissue culture flasks and the medium was change every third day until cells achieved 80% of confluence. BCH were used at passage 2 and hMSCs at passage 3. Prior to cell seeding scaffolds were sterilized with 70% (v/v) ethanol overnight and then rinsed three times in PBS, whereas surfaces were treated with ultraviolet (UV) light for 10 min to avoid the damage of the coating. Scaffolds and flat surfaces were then immersed for 4 hours in the medium appropriate for each cell type. For the scaffolds the seeding was performed by applying the cell suspension, with a concentration of $0.5\times10^6$ cells in 25 μL of medium (per scaffold). For surfaces the cell concentration was adjusted to $1.32\times10^4$ cells in 25 μL of medium (per glass coverslips). After cell attachment for 2 hours (37°C in a 5% $CO_2$), chondrocytes proliferation medium, MSCs proliferation medium or differentiation medium (DMEM, 2 mM glutamine (Gibco), 0.2 mM ASAC (Invitrogen), 100 μg.mL$^{-1}$ penicillin/Streptomycin (Invitrogen), 0.4 mM proline (Sigma-Aldrich), 100 μg.mL$^{-1}$ sodium pyruvate (Sigma-Aldrich) and 50 mg/mL insulin-Transferrin-selenite (ITS+premix, BD biosciences), 10 ng.mL$^{-1}$ TGFβ-3 (R&D systems) and 0.1 μM dexamethasone (Sigma-Aldrich)) was added.

## Cell viability

Cell viability and morphology were assessed with live/dead assay, MTT assay and SEM analysis. The scaffolds were cut in half in order to perform live/dead and MTT assays at 1, 3, 14 and 21 days. Scaffolds were further observed by SEM. For the surfaces the live dead assay was performed at 1, 3, 7, 14 and 21 days followed by SEM visualization. Medium was changed every third day to maintain an adequate supply of cell nutrients.

### Live/dead assay

To perform this assay the proliferation medium was aspirated from the wells in which the scaffolds and surfaces were deposited. The scaffolds and surfaces were then incubated with ethidium homodimer-1 (4 μM) and calcein-AM (2 μM) in PBS for 30 min at 37°C in a 5% $CO_2$atmosphere incubator. After 30 min the samples were immediately examined under an inverted fluorescent microscope (Nikon Eclipse E600) using Fluorescein isothiocyanate (FITC) and Texas Red Filter, as well as the NIS element-F.30 software.

### MTT assay

The scaffolds were incubated in 900 μL of proliferation medium and 100 μL of MTT solution (5 $mg.mL^{-1}$) per well for 2 h at 37°C in 5% $CO_2$. MTT staining images were captured using a stereomicroscope with colour camera (Nikon SMZ-10A) and the Qcapture software.

### Scanning electron microscopy observation

The structures with cells were fixed in formalin (10%) and dehydrated using serial concentrations of ethanol [70%, 80%, 90%, 96% and 100% (v/v), 30 min in each], before performing critical point drying (Balzers CPD 030). The samples were then coated with a conductive layer. The SEM observations were performed in a Philips XL 30 ESEM-FEG operated at 7.5–15 kV accelerating voltage.

### DNA quantification

Scaffolds seeded with BCH and hMSCs in differentiation medium at 1, 14 and 35 days were washed with PBS and frozen at −80°C before proteinase K (Sigma Aldrich) digestion. Then the scaffolds were digested with 1 mg/mL of proteinase K in tris (hydroxymethyl) aminomethane ethylenediaminetetraacetic (Tris\EDTA) buffer (pH = 7.6) containing 18.5 $\mu g.mL^{-1}$ idoacetamide and 1 μg/mL pepstatin

A (Sigma Aldrich) at 56°C for 20 hours. Quantification of total DNA in each sample was determined with CyQuant DNA kit according to manufacturer description (Molecular probes, Eugene, Orgeon, USA), using a spectrofluorometer (Victor3, Perkin-Elmer, USA) at an emission wavelength of 520 nm and an excitation wavelength of 480 nm.

## Histology

Haematoxylin & eosin (H&E) and alcian blue staining was used to analyse cell distribution and cartilage tissue formation, respectively. For histology analysis, scaffolds were fixed overnight in 10% formalin, and then dehydrated using sequential ethanol series [70%, 80%, 90%, 96%, and 100% (v/v), 30 min in each]. Once dehydrated, they were incubated in butanol overnight at 4°C and then in a solution of paraffin at 56°C for 12 hours. Sections of 4.5 μm were cut using a microtome (MicroM HM355S). After deparaffinization with xylene and rehydration using a graded ethanol series [from 100% to 70% (v/v)], the samples were stained using an automatic stainer (MicroM HMS740). For H&E staining samples were stained with haemotoxylin for 1 min and rinsed up to 6 min before being stained with eosin for 30 s. For alcian blue staining the samples were placed in alcian blue solution (0.5%, pH = 1) for 30 min and rinsed with tap water or distilled water for 4 min. In the last step nuclear fast red was added for 5 min before dehydration. Slides were assembled with resinous medium and mounted slides were examined under a light of Axioplan Imager Z1 microscope (Zeiss). Representative images were captured using a digital camera (AxioCAM MRCE) and treated using Axiovision software. Each assay was performed after 1, 14, 21 and 35 days of BCH and hMSCs culture, as well as with CHIT/CS scaffold as control.

## Statistical Analysis

The experiments were carried out in triplicate unlike otherwise specified. The results were presented as mean ± standard deviation (SD). Statistical analysis was performed using one way ANOVA followed by Turkey test (Graph Pad Prism 5.0 for Windows). Statistical significance was set to $p<0.05$ (*) and $p<0.01$ (**).

# RESULTS AND DISCUSSION

## CHT/CS film build-up

The build-up mechanism of the polymeric multilayered films made of CS and CHT was assessed in situ with QCM-D. This technique detected the adsorbed mass of polyelectrolytes and measured the viscoelastic properties of the surface [27], [28].

Figure 1A shows the build-up of 10 layers of CHT and CS in terms of variations on normalized frequency, ?$f_n/n$ (where n is the frequency overtone) and dissipation, ?$D_7$. As expected, the normalized frequency decreased with each CHT and CS solutions injection, reflecting the increased mass over the gold sensor. The increase of ?$D_7$ was due to the non-rigid adsorbed layer structure of the deposited film. During the washing step after the injection of each polyelectrolyte, the change of both ?$f_7/7$ and ?$D_7$ were relatively small compared to the total frequency variation associated to the adsorption of the respective polymer. This indicated a strong association of the layers on the surface of the crystal.

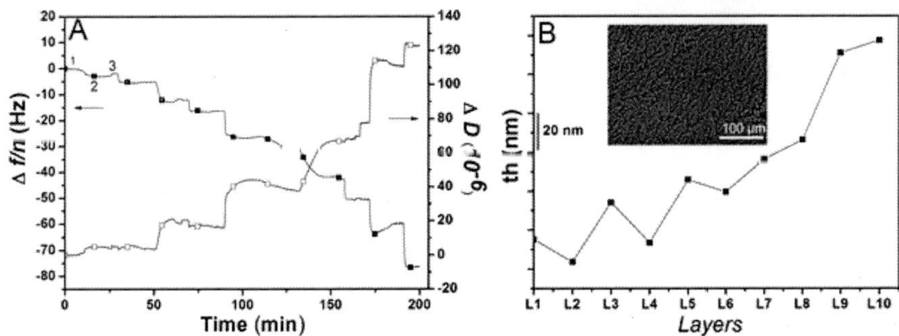


Figure 1. Build-up monitoring of the CHT/CS polyelectrolyte multi-layered using QCM for film constructed.

A) Normalized frequency (?$f_7$/7) and dissipation changes (?$D_7$) obtain at 35 MHz, 1) deposition of CHT, 2) washing step and 3) deposition of CS; B) Estimated thickness (th) evolution and SEM micrographs of the multilayer surface with 10 double layers (inset image).
doi:10.1371/journal.pone.0055451.g001

The combination of ?$f_7$/7 and ?$D_7$ gave information about the adsorbed amount and the variations of the viscoelastic properties [28]–[30]. The thickness of the film was estimated using the Voigt Model [23]. Figure 1B showed the thickness variation along the deposition of 10 layers. The results revealed a decrease of thickness from the first layer to the second one, which could be explained due to changes in water absorption [31]. The absorption of water was due to the presence of some groups in the polysaccharides (hydroxyl, carboxyl and sulphate groups) that interacted favourably with water molecules [31]. When the second layer was adsorbed the presence of opposite charge led to electrostatic interactions between them and the counterion-polymer. Consequently, water-polymer bonds were disrupted, resulting in an effective decrease of the hydrated film thickness [30]–[32]. The trend was observed during the first three pairs of layers. After the first three dL, this trend was no longer observed: there was an increase of thickness with the addition of CS. The SEM microphotography of the multilayered surface revealed a homogenous coating along the 2D flat surface (inset image ofFigure 1B). Moreover, the surface presented a rough texture and some granularity, with characteristic diameter sizes around 2 μm as measured by Image J.

The results obtained through QCM measurements and SEM demonstrated that CS could be successfully used with CHT to conceive a homogeneous viscoelastic polymeric self-assembled coating using the LbL approach.

## Multilayer surface

Using LbL methodology it was possible to produce surfaces with tuned properties [10]–[12].In this work, multilayers of CHT and CS were prepared on glass coverslips by using the LbL methodology, obtaining self-assembled films with 10 double layers.

## Cell behaviour in multilayers

In order to assess the cell viability in the surfaces, live/dead assay was performed (Figure 2A, 2B, 2D, 2E, 2G, 2H, 2J, 2K, 2M and 2N). The results showed a large amount of living cells and an increase in terms of cell number which results in cell confluence and in a continuous staining of calcein.

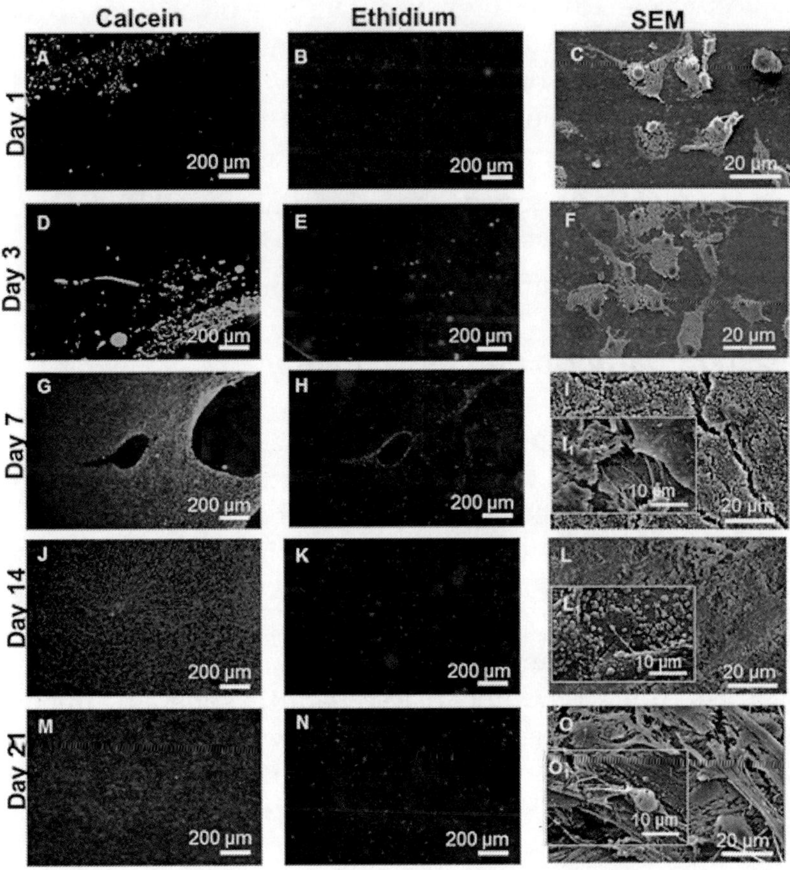


Figure 2. Live/dead assay and SEM micrographs of BCH seeded on glass coverslips coated with chitosan and chondroitin sulphate at day 1 (A, B, C), 3 (D, E, F), 7 (G, H, I), 14 (J, K, L) and 21 (M, N) of culture in proliferation medium.

doi:10.1371/journal.pone.0055451.g002

The cell adhesion/morphology was also studied using SEM (Figure 2C, 2F, 2I, 2L and 2O). The results revealed that the BCH were attached to the surface from the earliest time points onwards. Attachment, adhesion and spreading are the first phase of cell/material interaction and the quality of this stage influenced the capacity of cells to proliferate and differentiate itself on contact with an implant [33], [34]. With increasing culture time, cells started to spread out along the surface, losing their round phenotype which might occur due to the high proliferation of cells or to the 2D environment that leads to de-differentiation. As a result of significant cell proliferation most of the surface area was already covered with cells after 7 days of culture. At 14 days cellular confluence was achieved.

These results suggested the potential of CHT and CS as polyelectrolytes for the fabrication of the 3D nanostructure.

Nanostructured Scaffolds: physicochemical characterization

## Scaffold preparation and morphology

The use of bottom-up approaches to produce 3D porous structures is of particular interest to TE due to the hierarchical organization of the native tissues. It has been hypothesized that an interconnected 3D porous structure could be prepared combining LbL with leaching of free-packet paraffin spheres. A drop-wise addition method of PEMs over the 3D template formed by free-packet paraffin spheres was applied. This technique allows the formation of a 3D lattice arrangement from a randomly placed paraffin spheres. After coating the paraffin template was leached out and void spaces were created. Thus, the remaining material should be entirely composed by the CHT/CS multilayers (Figure 3A and Figure 3B).

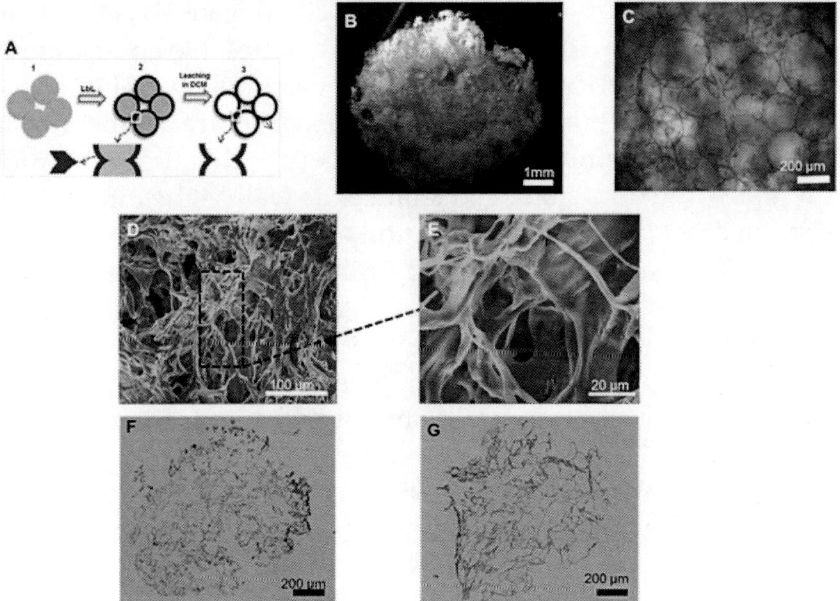


**Figure 3. Scaffold characterization.**

A) Production steps of scaffolds: LbL and leaching of free-packet paraffin spheres, B)Digital photograph of the scaffold after all the steps C) Optical Microscopy image of the scaffolds after the leaching of the core material, D, E) SEM micrographs of cross-sections (two different magnifications) and Histological cross-sections of the scaffolds after staining with alcian blue (F) and eosin (G).

doi:10.1371/journal.pone.0055451.g003

The morphology of the obtained scaffolds after the leaching was observed by optical microscopy. The results clearly revealed a bubble-like morphology with geometry and pore sizes consistent with the paraffin spheres used as the template (Figure 3C). The paraffin spheres used as porogen had a diameter of 200 µm which is appropriate for cartilage TE approaches, allowing the deposition of ECM and cell infiltration even after the commonly shrinking after the freeze drying process [22]. The interconnectivity should be assured by the existence of physical contact points between the neighbouring paraffin beads that will result in a passage point between the two pores after the leaching process (see red arrow in Figure 3A). Further structural information was obtained by SEM (Figure 3D and 3E). SEM images of freeze-dried scaffolds revealed a noticeable hollow imprint of the porous spherical wax template morphology. This concept allowed the production of highly porous structure with controlled pore size and interconnectivity. Consequently, this type of scaffolds should allow the diffusion of substances as well as the integration of cells, namely its infiltration, migration and distribution in the entire volume of the scaffold. The histological cross-sections of the freeze-dried scaffold stained by alcian blue (Figure 3F) and eosin (Figure 3G) showed a homogeneous distribution of the polysaccharides. Alcian blue stained chondroitin sulphate [35] and eosin chitosan due to the high ability of this polysaccharide to adsorb anionic dyes [36].

## Fourier transform infrared spectroscopy

FTIR measurements (Figure 4A) were performed on the scaffold produced, as well as on both CHT and CS powders in order to identify the presence of both polysaccharides in the entire specimen. The spectra of CHT and CS were very similar, as expected, reflecting the similarities in the chemical structure of both materials. As a result, they shared some common peaks around 3400 cm-1 corresponding to –OH and N-H bond stretching vibrations, and the peaks around 2900 cm-1 corresponding to C-H stretching. Between 1020 cm-1 and 1080 cm-1 the peaks associated with the stretching of C-O bonds could be observed also. Moreover, the amide groups appeared at 1648 cm-1 [37].

In the CHT spectrum the amine group bonds, characteristic of this polysaccharide, appeared at 1570 cm-1 [18]. The representative peak of chondroitin sulphate was detected at 1250 cm-1 corresponding to the stretching in the S = O bond ($SO_4^{2-}$) [37]. The spectrum of the scaffold showed globally the absorption peaks from both CHT and CS which was indicative of the presence of both raw materials in the final structure.

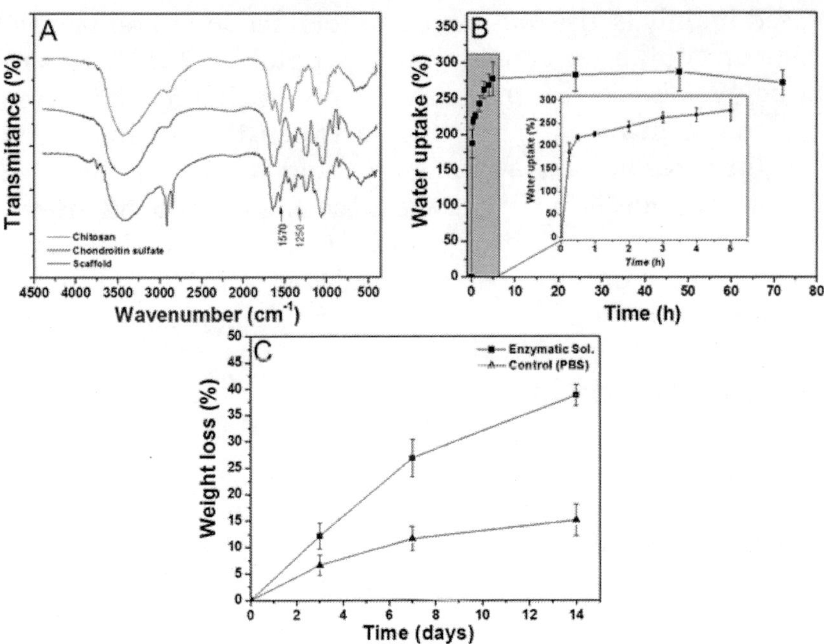


Figure 4. Physicochemical characterization of scaffolds.

A) FTIR measurements of CHT/CS scaffolds and pure polysaccharides (CHT and CS), B) Swelling test up to 3 days (The inset graphic expands the water uptake for the first 5 hours), C) Weight loss of the scaffolds in PBS (▲) and in an enzymatic solution at 37°C (▪).

doi:10.1371/journal.pone.0055451.g004

## Swelling ability

Water uptake is particularly important for implantable materials because it allows the diffusion and exchange of nutrients and waste through the entire scaffold; moreover the water uptake ability also influences the mechanical performance of the biomaterial [9]. The materials used in the scaffold have abundant hydrophilic groups, such as hydroxyl, amino, sulphate and carboxyl groups, which can promote the swollen state of the scaffold [38]–[40]. The swelling ability was evaluated by soaking scaffolds in PBS (pH 7.4) at 37°C for 3 days (Figure 4B). The results showed that the water uptake increased mainly in the first hour and then tended to remain stable, reaching an equilibrium after 5 h (water uptake = 280%). This result could be explained with the high density of charge that increased the difference in osmotic pressure between the scaffold network and medium, resulting in a swollen scaffold. Moreover, the swelling ability of cartilage is well known to be highly dependent on the binding of water to polar groups of GAGs, namely carboxylate and sulphate groups, on electrostatic repulsion of GAGs and entropic contributions resulting from the mixing of water and counterions [41].

## Enzymatic degradation

The biodegradability profile of scaffolds will dictate the changes in the structure that will occur upon the implantation. Enzymatic activity plays a fundamental role in the degradation of polysaccharides in vivo [42]. In vitro enzymatic degradation tests were performed with lysozyme and hyaluronidase solution and compared with weight loss in PBS (control). These two enzymes were chosen because they are present in the synovial fluid and they have as well the ability to cleave the polysaccharides used in this study [43], [44]. Lysozyme is able to degrade CHT and hyaluronidase has the ability to degrade both CHT and CS [44]–[47]. The weight loss as a function of time is presented in Figure 4C.

The results showed that the scaffolds degraded in the presence of the selected enzymes, showing weight losses of ca. 40% after 14 days. The degradation of scaffolds in the presence of the enzymatic solution was likely facilitated by the high porosity and interconnectivity of the structures allowing the easy access of the enzyme to their substrate. Moreover, the high hydrophilicity of scaffold (revealed by the high water uptake) could increase the interaction of scaffolds with the enzymatic solution, promoting the weight loss. The scaffolds placed in PBS also suffered some weight loss of ca. 15% after 14 days. In this case the weight loss could be the result of some disaggregation of the multilayered structure, as the polyelectrolytes were self-assembled through electrostatic interactions. The ions present in PBS may destabilize the structure and promote partial detachment between the macromolecules resulting in their release to the medium.

## Mechanical Properties

The viscoelastic/mechanical properties of an implantable device are fundamental for its performance *in vivo* [24]. Dynamical mechanical analysis (DMA) is an adequate non-destructive tool to characterize the mechanical and viscoelastic properties of polymeric materials [48],[49]. Since articular cartilage often is exposed to dynamic compression forces, DMA experiments were performed in a hydrated environment and at 37°C allowing the assessment of the mechanical properties of the scaffolds in more realistic conditions [24]. The storage modulus (E′) and loss factor (tan δ) as a function of frequency of the developed scaffolds, in the dry and wet state are presented in Figure 5. The results for the hydrated scaffold showed a slight increase in both E′ and tan δ with increasing frequency. In the dry state the values of E′ were about one order of magnitude higher when compared with the wet state. This was consistent with the high water uptake ability of the scaffolds and the plasticization effect of water molecules in such kind of polysaccharides increasing their molecular mobility and decreasing the stiffness of the material. A similar loss of the stiffness due to the effect of water was observed in CHT membranes [50]. In both cases no evident variation of E′ along the frequency axis were seen, indicating that no relaxation phenomena took place in the scaffolds within the time scale covered by the experiments. The tan δ of the dry sample decreased slightly with an increasing of frequency. However an opposite trend was observed when the samples were immersed in PBS. tan δ was higher in the wet samples, indicating that some dragging of entrapped water participated in energy loss for the hydrated structure [51]. The DMA results demonstrated viscoelastic behaviour of the scaffold which approached the viscoelastic nature of native cartilage [52]. Moreover, the E′ values obtained in the wet state (0.6–0.8 MPa, 0.01 – 10 Hz) were included in the range of mandibular condylar cartilage E′ values (0.1–1.5 MPa, 0.01 – 10 Hz ) [53].

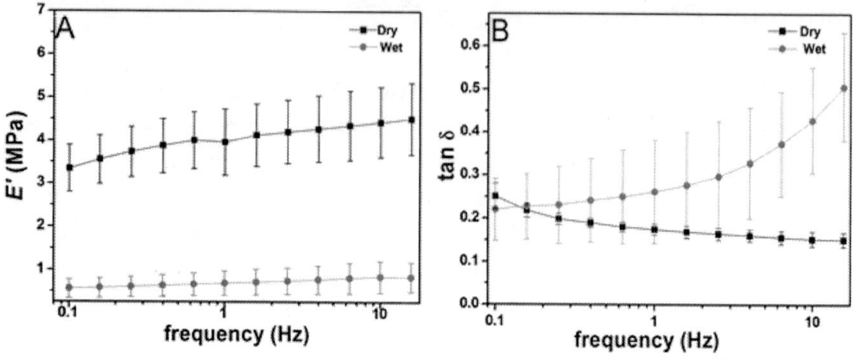


Figure 5. Variations of (A) Storage modulus (E′) and (b) loss factor (tanδ) of the CHT/CS scaffolds obtained by LbL methodology.

Experiments are reported for dry samples (▪) and hydrated samples in PBS at 37°C (•).

doi:10.1371/journal.pone.0055451.g005

Cell behaviour in nanostructured scaffolds

## CELL VIABILITY, ADHESION AND MORPHOLOGY

The cell viability tests with BCH (Figure 6A, 6D, 6G and 6J) and hMSCs (Figure 7A, 7D, 7G and 7J) showed evidence of cell attachment and a large amount of living cells (green)). This was consistent with the results obtained on flat surfaces.. After 1 day it was possible to see that the cells tended to aggregate. Furthermore, the results obtained with MTT assay for BCH (Figure 6B, 6E, 6H and 6K) and hMSCs (Figure 7B, 7E, 7H and 7K) suggested an increase in cell number and metabolic activity due to the increase in dark purple staining over time when compared with the control, the empty scaffold (Figure S1).

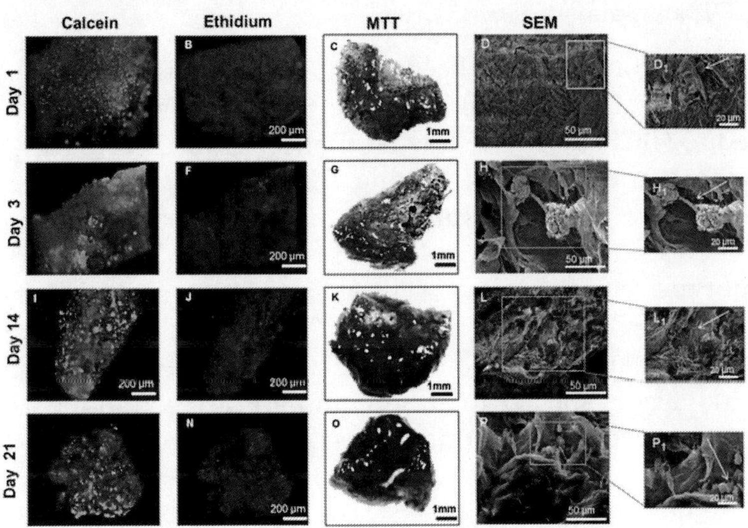


Figure 6. Live/dead assay, MTT assay and cross-section SEM micrographs of BCH seeded on scaffold at day 1(A, B, C), 3(D, E, F), 14 (G, H, I) and 21(J, K, L) of culture in proliferation medium.

doi:10.1371/journal.pone.0055451.g006

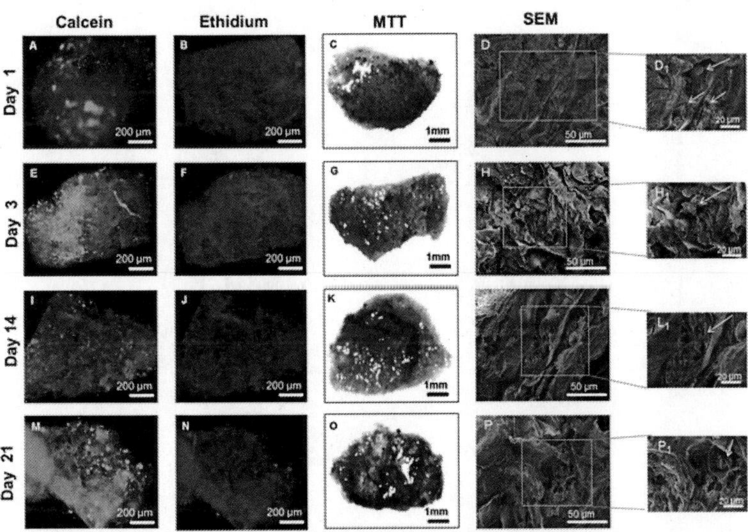


Figure 7. Live/dead assay, MTT assay and cross-section SEM micrographs of hMSCs seeded on scaffold at day 1(A, B, C), 3(D, E, F), 14 (G, H, I) and 21(J, K, L) of culture in proliferation medium.

doi:10.1371/journal.pone.0055451.g007

Cell adhesion and morphology was further studied by SEM using cross-sections of the scaffolds (Figure 6C, 6F, 6I and 6L). The results obtained for BCH at day 1 showed that cells attached to the surface, displaying a round shape. During the following culture time the adherent cells were more spread out along the scaffolds. The BCH presented a round shape in all time points which was an indication of phenotype retention and essential for matrix deposition [54]. The results for hMSCs (Figure 7C, 7F, 7I and 7L) revealed that the cells were attached to the surface and presented a more stretched morphology. After 1 day, cells started to adhere and an increase in terms of cell migration occurred in the inner areas of the scaffolds (t = 14 days).

## CELL PROLIFERATION

Cell proliferation in differentiation medium was evaluated using a DNA assay (Figure 8). The result obtained for the two types of cell showed that the number of both types of cells increased with increasing culture time.

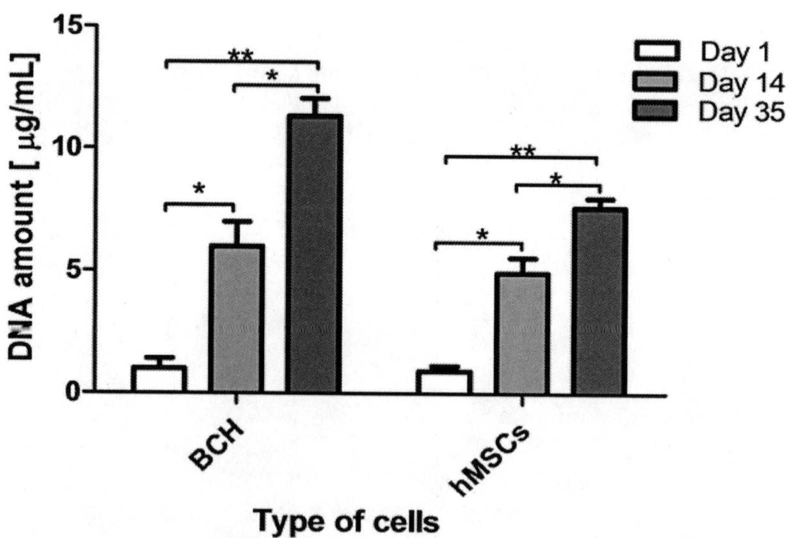


Figure 8. DNA assay on the scaffolds seeded with BCH and hMSCs in differentiation medium.

Between the first day of culture and after 35 days there were significant differences in the amount of BCH and hMSCs, indicating that the cells continued proliferation even after long time culture which corroborated the results of cell viability tests.

Significant differences between each cell type at different time points were found for $p<0.05$(*) and $p<0.01$(**).

doi:10.1371/journal.pone.0055451.g008

## Histology

Cell distribution and matrix production in differentiation medium was evaluated using histology cross-sections stained with H&E and alcian blue (Figure 9).

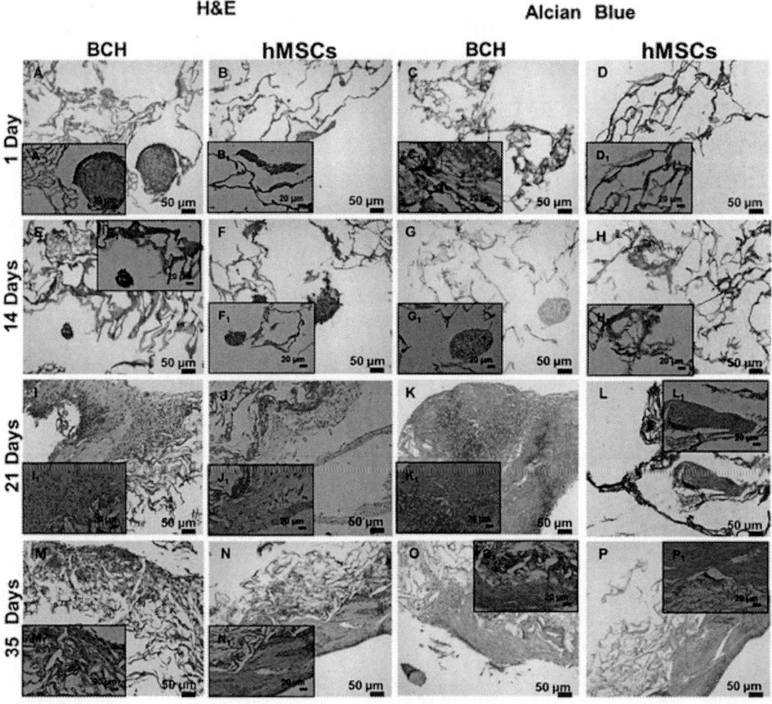


Figure 9. Histological cross-sections of scaffolds seeded with BCH and hMSCs stained by H&E and Alcian blue at different days of culture in differentiation medium.

doi:10.1371/journal.pone.0055451.g009

The H&E staining of scaffold seeded with BCH showed the round morphology of cells. Moreover, over time the abundance of cells per section was increased in support of cell proliferation. At day 1, cells started to attach to the walls forming small aggregates. At day 14 the size of BCH agglomerates increased. During the following weeks, the cells presented a higher dispersion and distribution in the scaffolds. Sulphated GAGs, indicating new cartilage matrix formation, were stained by alcian blue. CS, which gave as well a positive staining for GAGs, can be distinguished from newly deposited matrix by comparing alcian blue staining at day 1 with later time points. Secretion of GAGs by BCH was first observed from 14 days of culture. GAG production increased during subsequent weeks. Lacunae formation was also seen in the matrix surrounding BCH, namely at day 21 and 35. The maintenance of chondrogenic phenotype is indicated by the lacunae formation.

In scaffolds seeded with hMSCs it was possible to see some agglomerated cells at day 1 and after 2 weeks the size of these agglomerates increased (Figure 9). During the following time points the hMSCs were more spread throughout the scaffold. GAG deposition was also assessed and at day 14 a small amount of deposition could be seen. The amount of deposition increased during the next weeks. The deposition of GAGs by hMSCs indicated the chondrogenic differentiation of these cells. The driving force for differentiation in this assay was TGF-β, although a role of CS in chondrogenic differentiation cannot be excluded [35].

## CONCLUSIONS

Flat CHT/CS PEMs prepared using LbL elicit a positive effect on BCH cells, allowing its attachment and proliferation. It was possible to use LbL combined with spherical template leaching to produce an innovative 3D nanostructured constructs of CHT/CS with high porosity and interconnectivity, just composed by self-assembled multilayers of these polyelectrolytes. Both BCH and hMSCs could adhere and proliferate in these scaffolds, Secretion of GAGs was observed in BCH and hMSCs upon culture in chondrogenic differentiation medium, indicating that the chondrogenic phenotype was maintained and hMSCs differentiation was successfully induced. Our results suggest that nanostructured scaffolds of chitosan and chondroitin sulphate obtained by LbL technology could have potential use in TE approaches for cartilage namely for matrix blunt and partial thickness/chondral defects. The scaffold would be implant after cell culture *in vitro* in order to increase the production of matrix that will start under static conditions after 14 days. The use of low oxygen tension, mechanical stimulation could accelerate this process [55]. As future work we envisage the production of scaffolds with an increase in terms of mechanical properties by incorporation of fillers, such as nanotubes or using crosslinked PEMs that will also reduce its degradation rate. The ability of these PEMs to sustain delivery of growth factors such as TGF-β could be another way to improve its performance. These improvements will open new horizons for clinical application in the field of cartilage tissue engineering.

## Author Contributions

Conceived and designed the experiments: JMS NG RC PS MK JFM. Performed the experiments: JMS NG RC PS. Analyzed the data: JMS NG RC PS MK JFM. Contributed reagents/materials/analysis tools: JMS NG RC PS CAVB RLR MK JFM. Wrote the paper: JMS NG RC PS MK JFM.

## REFERENCES

1. Fan H, Hu Y, Zhang C, Li X, Lv R, et al. (2006) Cartilage regeneration using mesenchymal stem cells and a PLGA-gelatin/chondroitin/hyaluronate hybrid scaffold. Biomaterials 27: 4573–4580. doi: 10.1016/j.biomaterials.2006.04.013
2. LeBaron RG, Athanasiou KA (2000) Ex vivo synthesis of articular cartilage. Biomaterials 21: 2575–2587. doi: 10.1016/S0142-9612(00)00125-3
3. Heng BC, Cao T, Lee EH (2004) Directing Stem Cell Differentiation into the Chondrogenic Lineage In Vitro. STEM CELLS 22: 1152–1167. doi: 10.1634/stemcells.2004-0062
4. Mano JF, Reis RL (2007) Osteochondral defects: present situation and tissue engineering approaches. Journal of Tissue Engineering and

Regenerative Medicine 1: 261–273. doi: 10.1002/term.37

5. Temenoff JS, Mikos AG (2000) Review: tissue engineering for regeneration of articular cartilage. Biomaterials 21: 431–440. doi: 10.1016/S0142-9612(99)00213-6
6. Chung C, Burdick JA (2008) Engineering cartilage tissue. Advanced Drug Delivery Reviews 60: 243–262. doi: 10.1016/j.addr.2007.08.027
7. Chen FH, Rousche KT, Tuan RS (2006) Technology Insight: adult stem cells in cartilage regeneration and tissue engineering. Nat Clin Pract Rheum 2: 373–382. doi: 10.1038/ncprheum0216
8. Lee S-H, Shin H (2007) Matrices and scaffolds for delivery of bioactive molecules in bone and cartilage tissue engineering. Advanced Drug Delivery Reviews 59: 339–359. doi: 10.1016/j.addr.2007.03.016
9. Puppi D, Chiellini F, Piras AM, Chiellini E (2010) Polymeric materials for bone and cartilage repair. Progress in Polymer Science 35: 403–440. doi: 10.1016/j.progpolymsci.2010.01.006
10. Boudou T, Crouzier T, Ren K, Blin G, Picart C (2010) Multiple Functionalities of Polyelectrolyte Multilayer Films: New Biomedical Applications. Advanced Materials 22: 441–467. doi: 10.1002/adma.200901327
11. Detzel C, Larkin A, Rajagopalan P (2011) Polyelectrolyte Multilayers in Tissue Engineering. Tissue Engineering Part B: Reviews 17: 101–113. doi: 10.1089/ten.TEB.2010.0548
12. Tang Z, Wang Y, Podsiadlo P, Kotov NA (2006) Biomedical Applications of Layer-by-Layer Assembly: From Biomimetics to Tissue Engineering. Advanced Materials 18: 3203–3224. doi: 10.1016/j.progpolymsci.2010.01.006
13. Almodo?var J, Place LW, Gogolski J, Erickson K, Kipper MJ (2011) Layer-by-Layer Assembly of Polysaccharide-Based Polyelectrolyte Multilayers: A Spectroscopic Study of Hydrophilicity, Composition, and Ion Pairing. Biomacromolecules 12: 2755–2765. doi: 10.1021/bm200519y
14. Sher P, Custódio CA, Mano JF (2010) Layer-By-Layer Technique for Producing Porous Nanostructured 3D Constructs Using Moldable Freeform Assembly of Spherical Templates. Small 6: 2644–2648. doi: 10.1002/smll.201001066
15. Miranda E, Silva T, Reis R, Mano J (2011) Nanostructured Natural-Based Polyelectrolyte Multilayers to Agglomerate Chitosan Particles into Scaffolds for Tissue Engineering. Tissue Engineering Part A 17: 2663–2674. doi: 10.1089/ten.tea.2010.0635
16. Vinatier C, Mrugala D, Jorgensen C, Guicheux J, Noël D (2009)

Cartilage engineering: a crucial combination of cells, biomaterials and biofactors. Trends in Biotechnology 27: 307–314. doi: 10.1016/j.tibtech.2009.02.005

17. Francis Suh JK, Matthew HWT (2000) Application of chitosan-based polysaccharide biomaterials in cartilage tissue engineering: a review. Biomaterials 21: 2589–2598. doi: 10.1016/S0142-9612(00)00126-5
18. Neves SC, Moreira Teixeira LS, Moroni L, Reis RL, Van Blitterswijk CA, et al. (2011) Chitosan/Poly(?-caprolactone) blend scaffolds for cartilage repair. Biomaterials 32: 1068–1079. doi: 10.1016/j.biomaterials.2010.09.073
19. Kubo M, Ando K, Mimura T, Matsusue Y, Mori K (2009) Chondroitin sulfate for the treatment of hip and knee osteoarthritis: Current status and future trends. Life Sciences 85: 477–483. doi: 10.1016/j.lfs.2009.08.005
20. Pieper JS, Oosterhof A, Dijkstra PJ, Veerkamp JH, van Kuppevelt TH (1999) Preparation and characterization of porous crosslinked collagenous matrices containing bioavailable chondroitin sulphate. Biomaterials 20: 847–858. doi: 10.1016/S0142-9612(98)00240-3
21. Gong Y, Zhu Y, Liu Y, Ma Z, Gao C, et al. (2007) Layer-by-layer assembly of chondroitin sulfate and collagen on aminolyzed poly(l-lactic acid) porous scaffolds to enhance their chondrogenesis. Acta Biomaterialia 3: 677–685. doi: 10.1016/j.actbio.2007.04.007
22. Liu Y, He T, Gao C (2005) Surface modification of poly(ethylene terephthalate) via hydrolysis and layer-by-layer assembly of chitosan and chondroitin sulfate to construct cytocompatible layer for human endothelial cells. Colloids and Surfaces B: Biointerfaces 46: 117–126. doi: 10.1016/j.colsurfb.2005.09.005
23. Voinova MV, Rodahl M, Jonson M, Kasemo B (1999) Viscoelastic Acoustic Response of Layered Polymer Films at Fluid-Solid Interfaces: Continuum Mechanics Approach. Physica Scripta 59: 391. doi: 10.1238/physica.regular.059a00391
24. Sobral JM, Caridade SG, Sousa RA, Mano JF, Reis RL (2011) Three-dimensional plotted scaffolds with controlled pore size gradients: Effect of scaffold geometry on mechanical performance and cell seeding efficiency. Acta Biomaterialia 7: 1009–1018. doi: 10.1016/j.actbio.2010.11.003
25. Correia C, MoreiraTeixeira L, Moroni L, Reis R, van Blitterswijk C, et al.. (2011) Chitosan Scaffolds Containing Hyaluronic Acid for Cartilage Tissue Engineering. Tissue Engineering Part C: Methods: 110308075242061.
26. Wu L, Leijten J, Georgi N, Post J, van Blitterswijk C, et al. (2011)

Trophic Effects of Mesenchymal Stem Cells Increase Chondrocyte Proliferation and Matrix Formation. Tissue Engineering Part A 17: 1425–1436. doi: 10.1089/ten.TEA.2010.0517

27. Costa RR, Custódio CA, Arias FJ, Rodríguez-Cabello JC, Mano JF (2011) Layer-by-Layer Assembly of Chitosan and Recombinant Biopolymers into Biomimetic Coatings with Multiple Stimuli-Responsive Properties. Small 7: 2640–2649. doi: 10.1002/smll.201100875

28. Costa RR, Custódio CA, Testera AM, Arias FJ, Rodríguez-Cabello JC, et al. (2009) Stimuli-Responsive Thin Coatings Using Elastin-Like Polymers for Biomedical Applications. Advanced Functional Materials 19: 3210–3218. doi: 10.1002/adfm.200900568

29. Alves NM, Picart C, Mano JF (2009) Self Assembling and Crosslinking of Polyelectrolyte Multilayer Films of Chitosan and Alginate Studied by QCM and IR Spectroscopy. Macromolecular Bioscience 9: 776–785. doi: 10.1002/mabi.200800336

30. Martins GV, Mano JF, Alves NM (2010) Nanostructured self-assembled films containing chitosan fabricated at neutral pH. Carbohydrate Polymers 80: 570–573. doi: 10.1002/adfm.200900568

31. Crouzier T, Boudou T, Picart C (2010) Polysaccharide-based polyelectrolyte multilayers. Current Opinion in Colloid & Interface Science 15: 417–426. doi: 10.1002/adfm.200900568

32. Köhler K, Shchukin DG, Möhwald H, Sukhorukov GB (2005) Thermal Behavior of Polyelectrolyte Multilayer Microcapsules. 1. The Effect of Odd and Even Layer Number. The Journal of Physical Chemistry B 109: 18250–18259. doi: 10.1021/jp052208i

33. Anselme K (2000) Osteoblast adhesion on biomaterials. Biomaterials 21: 667–681. doi: 10.1016/S0142-9612(99)00242-2

34. Kirkpatrick C, Dekker A (1992) Quantitative evaluation of cell interaction with biomaterials in vitro. Advanced biomaterials 10: 31–42.

35. Chen W-C, Yao C-L, Chu IM, Wei Y-H (2011) Compare the effects of chondrogenesis by culture of human mesenchymal stem cells with various type of the chondroitin sulfate C. Journal of Bioscience and Bioengineering. 111: 226–231.

36. Chatterjee S, Chatterjee S, Chatterjee BP, Das AR, Guha AK (2005) Adsorption of a model anionic dye, eosin Y, from aqueous solution by chitosan hydrobeads. Journal of Colloid and Interface Science 288: 30–35. doi: 10.1016/j.jcis.2005.02.055

37. Sui W, Huang L, Wang J, Bo Q (2008) Preparation and properties of chitosan chondroitin sulfate complex microcapsules. Colloids and

Surfaces B: Biointerfaces 65: 69–73.

38. Fajardo AR, Piai JF, Rubira AF, Muniz EC (2010) Time- and pH-dependent self-rearrangement of a swollen polymer network based on polyelectrolytes complexes of chitosan/chondroitin sulfate. Carbohydrate Polymers 80: 934–943. doi: 10.1016/j.carbpol.2010.01.009
39. Piai JF, Rubira AF, Muniz EC (2009) Self-assembly of a swollen chitosan/chondroitin sulfate hydrogel by outward diffusion of the chondroitin sulfate chains. Acta Biomaterialia 5: 2601–2609. doi: 10.1016/j.actbio.2009.03.035
40. Varghese JM, Ismail YA, Lee CK, Shin KM, Shin MK, et al. (2008) Thermoresponsive hydrogels based on poly(N-isopropylacrylamide)/ chondroitin sulfate. Sensors and Actuators B: Chemical 135: 336–341. doi: 10.1016/j.carbpol.2010.01.009
41. Servaty R, Schiller J, Binder H, Arnold K (2001) Hydration of polymeric components of cartilage – an infrared spectroscopic study on hyaluronic acid and chondroitin sulfate. International Journal of Biological Macromolecules 28: 121–127. doi: 10.1016/S0141-8130(00)00161-6
42. Dietmar WH (2000) Scaffolds in tissue engineering bone and cartilage. Biomaterials 21: 2529–2543. doi: 10.1016/S0142-9612(00)00121-6
43. Fraser JRE, Laurent TC, Laurent UBG (1997) Hyaluronan: its nature, distribution, functions and turnover. Journal of Internal Medicine 242: 27–33. doi: 10.1046/j.1365-2796.1997.00170.x
44. Martins AM, Santos MI, Azevedo HS, Malafaya PB, Reis RL (2008) Natural origin scaffolds with in situ pore forming capability for bone tissue engineering applications. Acta Biomaterialia 4: 1637–1645. doi: 10.1016/j.actbio.2008.06.004
45. Jin R, Moreira Teixeira LS, Krouwels A, Dijkstra PJ, van Blitterswijk CA, et al. (2010) Synthesis and characterization of hyaluronic acid-poly(ethylene glycol) hydrogels via Michael addition: An injectable biomaterial for cartilage repair. Acta Biomaterialia 6: 1968–1977. doi: 10.1016/j.actbio.2009.12.024
46. Mao JS, liu HF, Yin YJ, Yao KD (2003) The properties of chitosan-gelatin membranes and scaffolds modified with hyaluronic acid by different methods. Biomaterials 24: 1621–1629. doi: 10.1016/S0142-9612(02)00549-5
47. Menzel EJ, Farr C (1998) Hyaluronidase and its substrate hyaluronan: biochemistry, biological activities and therapeutic uses. Cancer Letters 131: 3–11. doi: 10.1016/S0304-3835(98)00195-5
48. Yan L-P, Oliveira JM, Oliveira AL, Caridade SG, Mano JF, et al.

(2012) Macro/microporous silk fibroin scaffolds with potential for articular cartilage and meniscus tissue engineering applications. Acta Biomaterialia 8: 289–301. doi: 10.1016/j.actbio.2011.09.037

49. Yan L-P, Wang Y-J, Ren L, Wu G, Caridade SG, et al. (2010) Genipin-cross-linked collagen/chitosan biomimetic scaffolds for articular cartilage tissue engineering applications. Journal of Biomedical Materials Research Part A 95A: 465–475. doi: 10.1002/jbm.a.32869
50. Mano JF (2008) Viscoelastic Properties of Chitosan with Different Hydration Degrees as Studied by Dynamic Mechanical Analysis. Macromolecular Bioscience 8: 69–76. doi: 10.1002/mabi.200700139
51. Ghosh S, Gutierrez V, Fernández C, Rodriguez-Perez MA, Viana JC, et al. (2008) Dynamic mechanical behavior of starch-based scaffolds in dry and physiologically simulated conditions: Effect of porosity and pore size. Acta Biomaterialia 4: 950–959. doi: 10.1016/j.actbio.2008.02.001
52. Moutos FT, Freed LE, Guilak F (2007) A biomimetic three-dimensional woven composite scaffold for functional tissue engineering of cartilage. Nat Mater 6: 162–167. doi: 10.1038/nmat1822
53. Franke O, Göken M, Meyers MA, Durst K, Hodge AM (2011) Dynamic nanoindentation of articular porcine cartilage. Materials Science and Engineering: C 31: 789–795. doi: 10.1016/j.msec.2010.12.005
54. Von Der Mark K, Gauss V, Von Der Mark H, Muller P (1977) Relationship between cell shape and type of collagen synthesised as chondrocytes lose their cartilage phenotype in culture. Nature 267: 531–532. doi: 10.1038/267531a0
55. Albrecht C, Tichy B, Nürnberger S, Hosiner S, Zak L, et al. (2011) Gene expression and cell differentiation in matrix-associated chondrocyte transplantation grafts: a comparative study. Osteoarthritis and Cartilage 19: 1219–1227. doi: 10.1016/j.joca.2011.07.004

# Chapter 4

# FABRICATION OF SMART CHEMICAL SENSORS BASED ON TRANSITION-DOPED-SEMICONDUCTOR NANOSTRUCTURE MATERIALS WITH μ-CHIPS

Mohammed M. Rahman[1,2*], Sher Bahadar Khan[1,2,] Abdullah M. Asiri[1,2]

[1]Center of Excellence for Advanced Materials Research, King Abdulaziz University, Jeddah, Saudi Arabia

[2]Chemistry Department, Faculty of Science, King Abdulaziz University, Jeddah, Saudi Arabia

## ABSTRACT

Transition metal doped semiconductor nanostructure materials ($Sb_2O_3$ doped ZnO microflowers, MFs) are deposited onto tiny μ-chip (surface area, ~0.02217 $cm^2$) to fabricate a smart chemical sensor for toxic ethanol in phosphate buffer solution (0.1 M PBS). The fabricated chemi-sensor is also exhibited higher sensitivity, large-dynamic concentration ranges, long-term stability, and improved electrochemical performances towards ethanol. The calibration plot is linear ($r^2$ = 0.9989) over the large ethanol concentration ranges (0.17 mM to 0.85 M). The sensitivity and detection limit is ~5.845 $\mu Acm^{-2}mM^{-1}$ and ~0.11±0.02 mM (signal-to-noise ratio, at a SNR of 3) respectively. Here, doped MFs are prepared by a wet-chemical process using reducing agents in alkaline medium, which characterized by UV/vis., FT-IR, Raman, X-ray photoelectron spectroscopy (XPS), powder X-ray diffraction (XRD), and field-emission scanning electron microscopy (FE-SEM) etc. The fabricated ethanol chemical sensor using $Sb_2O_3$-ZnO MFs is simple, reliable, low-sample volume (<70.0 μL), easy of integration, high sensitivity, and excellent stability for the fabrication of efficient I–V sensors on μ-chips.

## INTRODUCTION

The development of chemi-sensors has been the subject of considerable interest in recent years based on μ-chips using specific matrixes. Such modification on μ-chip offers great attention to chemical detection that included high-sensitivity, inherent-miniaturization, low-cost, high selectivity, low-sample volume, independence of sample turbidity or optical-path length, minimal-power demands, and high bio-compatibility with advanced micro-fabrication technologies. The development of electrochemical chemi-sensors for sensitive detection of toxic chemicals is generally required innovative approaches that coupled with various modification/amplification procedures onto electro-active substrates. Semiconductor codoped materials have attracted much interest because of their unique properties and potential applications in all areas of advance science and technological fields [1].

The simplest synthetic route for doped-materials is possibly self-aggregation, in which ordered doped aggregates are prepared in economical approaches [2]. Although, it is still a big-challenge to develop simple and economical route for microstructures semiconductor codoped metal oxide with designed chemical components and controlled morphologies, which is strongly influenced the optical and electrical properties of doped nanomaterials [3]. The significance of safety for human-beings and environments has been considered with great attention in doped semiconductor chemi-sensors for toxic chemical detection (toxic level in human blood, >0.10% or 2.2 mM) by reliable methods using μ-chips [4], [5]. Semiconductor micro-structure materials are very sensitive due to their particle-size and high-active surface-area as compared to the transition materials in micro-ranges. Nanoscale materials composed micro-structures have also displayed a huge-deal of consideration due to their promising properties such as large active-surface area, high-stability, quantum confinement, high-porosity, and permeability (meso-porous nature), which is directly dependent on the shape and size of the microcrystals [6], [7].

During the last two decades, semiconductor nanomaterials have been received significant attention due to their electronic, magnetic, electrical, optoelectronic, mechanical properties, and their prospective applications in nanotechnology fields. Doped materials might be a promising candidate because of their high-specific surface-area, low-resistance, fascinating electrochemical, and optical properties [8], [9]. Solution–liquid–solid mechanism [10], vapor–solid mechanism [11], and oxide-assisted growth mechanism [12] have also been adopted to prepare various antimony oxide nanostructure materials. Recently, antimony oxide nanostructure have been also prepared using micro-emulsion [13] and templating CNTs [14] by several groups, however, nanosheets composed microstructure of $Sb_2O_3$-doped ZnO MFs have never reported. In this report, it is displayed an alternative approach to the synthesis of $Sb_2O_3$-ZnO MFs with diameters of several nanometers sheet composed several micrometers flowers by a wet-chemical process. Here, semiconductor zinc oxide (ZnO, band-gap ~3.4 eV, II–VI compound, binding energy ~60 meV, excellent acoustic wave, and n-type semiconductor) has been recognized as a promising host material at room conditions, which displayed a wurtzite type structure with number of periodic planes with hexagonally-coordinated O and Zn atoms piled along the c-axis [15].

For outstanding and extraordinary properties of ZnO, it is used for flexible applications in piezo-electric chips, opto-electronics, photo-catalytic, solar cells, transparent thin- film transistors, bio- and chemi-sensors, spintronics, light-emitters, electronics, catalysis, and so forth [16]–[18]. For exotic and flexible properties including bio-compatibility, non-toxicity, chemical and photo-chemical stability, high-specific surface area, optical and electro-chemical behaviors, and high-electron communication characteristics, the transition-doped semiconductor nanomaterials presents itself as one of the most promising materials for the development and fabrication of efficient chemi-sensors [19], [20]. Recently, the extensive progresses have been explored on ZnO-based nanomaterials synthesis by a wet-chemical and conventional techniques [21]. Zinc oxide nanostructure displayed attractive applications, such as transistors, UV photo-detectors, gas-sensors, field-emission electron sources, nano-wires and nano-lasers, nano-scale power generators, and many other functional devices [22]–[27].

Advances in nanotechnology for innovative chemi-sensors, nanomaterials embedded µ-chips have been regulating a key-task in the fabrication and improvement of very precise, perceptive, accurate, sensitive, and consistent sensors. The exploration for even tiny chips accomplished in nano-level imaging and controlling of doped nanomaterial for biological, chemical, pathological samples, chemi-sensor has recently expanded the spotlight of awareness of scientist mainly for control monitoring due to the amplifying essential for environmental safety and health monitoring [28], [29]. Transition-doped semiconductor metal oxides are the model materials for sensing due to high-active surface areas and extensively employed as sensor for the detection, recognition, and quantification of various toxic pollutants and hazardous chemicals [30]–[34]. In presence of ethanol, it causes damage of brain and specific diseases of stomach, liver, and erythrocyte. Therefore it is important and big challenge to sense ethanol efficiently and shield the human health from dangerous diseases and safe the environment using electro-analytical methods [35]. Recently, a large amount of undoped metal oxides is considered as chemical sensors for the detection of different hazardous pollutants and toxic chemicals [36], [37]. Therefore $Sb_2O_3$-ZnO MFs have been offered as a mediator to detect and quantify the ethanol chemicals in liquid phase. The main attention of the present investigation is to fabricate and develop a highly sensitive chemi-sensor (especially ethanol) for detecting and quantifying hazardous pollutants using $Sb_2O_3$-ZnO MFs on µ-chips. Hence a well-crystalline $Sb_2O_3$-ZnO MFs were prepared by a wet-chemical process and totally characterized by using UV/visible, FT-IR spectroscopy, XRD, FESEM, and XPS analysis etc.

In this contribution, it is prepared by a wet-chemical process to arrange as-prepared $Sb_2O_3$doped ZnO MFs with practically controlled flower-shape structures, which exposed a control-

morphological development in microstructure materials and potential chemi-sensor applications using tiny μ-chips. With most of the significant properties of the doped material, there have been more and more attention focused to explore the doped counterparts. For semiconductor materials, doping is an exceptional function to improve significantly the optical and electrical properties, which enhances the development of electronic and opto-electronic devices. Codoped nanosheets composed $Sb_2O_3$-ZnO MFs are used to fabricate by a simple hydrothermal method on a tiny μ-chip surfaces, which allows very sensitive transduction of liquid/surface interactions and measured the chemical sensing performance considering ethanol at ambient conditions. To best of our knowledge, this is the first report for detection of toxic ethanol with as-grown $Sb_2O_3$-ZnO MFs onto tiny μ-chips using reliable I-V method in short response time.

## EXPERIMENTAL SECTIONS

### Materials and Methods

Zinc chloride, butyl carbitol acetate, antimony chloride, ethyl acetate, ammonia solution (25%), and all other chemicals were in analytical grade and purchased from Sigma-Aldrich Company. They were used without further purification. The $\lambda_{max}$ (289.0 nm) of as-grown $Sb_2O_3$ doped ZnO MFs was executed using UV/visible spectroscopy Lamda-950, Perkin Elmer, Germany. FT-IR spectra of MFs were measured on a spectrum-100 FT-IR spectrophotometer in the mid-IR range purchased from Bruker (ALPHA, USA). The XPS measurements of MFs were executed on a Thermo Scientific K-Alpha KA1066 spectrometer (Germany). Monochromatic $AlK_\alpha$ x-ray radiation sources were used as excitation sources, where beam-spot size was kept in 300.0 μm. The spectrum was recorded in the fixed analyzer transmission mode (pass energy, ~200.0 eV), where the sample scanning of the spectra was performed less $10^{-8}$ Torr. Morphology, size, elemental, and structure evaluation of as-grown MFs were recorded on FE-SEM instrument from JEOL (JSM-7600F, Japan). The powder XRD patterns of MFs were recorded by X-ray diffractometer from PANalytical diffractometer equipped with $Cu\text{-}K_{\alpha 1}$ radiation ($\lambda$ = 1.5406 nm). The generator voltage (~45.0

kV) and generator current (~40.0 mA) were applied for the XRD measurement. Raman spectrometer was used to measure the Raman shift of as-grown MFs using radiation source ($Ar^+$ laser line, $\lambda$; 513.4 nm), which was purchased from Perkin Elmer (Raman station 400, Perkin Elmer, Germany). I–V technique (two electrodes composed on μ-chip) is measured by using Keithley-Electrometer from USA.

**Preparation and growth mechanism of $Sb_2O_3$ doped ZnO MFs**

Large-scale synthesis of $Sb_2O_3$-ZnO MFs was prepared by a wet-chemical process at low-temperature using zinc chloride ($ZnCl_2$), antimony chloride ($SbCl_3$), and ammonium hydroxide ($NH_4OH$). In a usual reaction process, 0.1 M $ZnCl_2$ dissolved in 50.0 ml deionized (DI) water mixed with 50.0 ml DI solution of 0.1 M $SbCl_3$ and 50.0 ml of 0.1 M urea under continuous stirring. The pH of the solution was adjusted to 9.7 by addition of $NH_4OH$ and resulting mixture was shacked and stirred continuously for 30.0 minutes at room conditions. After stirring, the solution mixture was then put into conical flux and heat-up at 160°C for 12.0 hours. The temperature of solution was controlled manually throughout the reaction process at 90.0°C. After heating the reactant mixtures, the flux was kept for cooling at room conditions until reached to room temperature. The final codoped products were obtained, which was washed thoroughly with DI water, ethanol, and acetone for several times subsequently and dried at room-temperature for structural, elemental, morphological, and optical characterizations. The growth mechanism of the $Sb_2O_3$-ZnO nanostructure materials can be explained on the basis of chemical reactions and nucleation as well as growth of doped nanocrystals. The probable reaction mechanisms are presented here for attaining the codoped nanomaterial oxides in below.

$$\mathbf{NH_4OH_{(aq)} \rightarrow 2NH_4^+ (aq) + OH^-_{(aq} \quad (i)}$$

$$\mathbf{SbCl_3 + 3NH_4OH_{(aq)} + H_2O_{(aq)} \rightarrow Sb(OH)_{3aq)} + 3NH_4Cl + 2H_2O_{(aq)} \quad (ii)}$$

$$\mathbf{ZnCl_{2(s)} + 2NH_4OH_{(aq)} \rightarrow Zn(OH)_{2(aq)} + 2NH_4{}^+_{(aq)} + 2Cl^-_{(aq)} \quad (iii)}$$

$$\mathbf{Zn(OH)_{2(aq)} + 4NH_4OH_{(aq)} + Sb(OH)_{3(aq)} \rightarrow Sb_2O_3.ZnO_{(s)} + 4NH_4^+ + 2Cl^- + H_2O \quad (iv)}$$

The reaction is forwarded slowly according to the proposed equation (i) to equation (iii). During preparation, the pH value of the reaction medium plays an important responsibility in the doped nano-material oxide formation. At a particular pH, when $SbCl_3$ is hydrolyzed with ammonia solution, antimony hydroxide is formed instantly according to the equation (ii). During the whole synthesis route, $NH_4OH$ operates a pH buffer to control the pH value of the solution and slow contribute of hydroxyl ions ($OH^-$). When the concentrations of the $Sb^{3+}$ and $OH^-$ ions are achieved above in critical value, the precipitation of $Sb_2O_3$ nuclei begin to start. As there is high concentration of $Zn^{2+}$ ions in the solution [according to the reactions (iii)], the nucleation of $Sb_2O_3$ crystals become slower due to the lower activation energy barrier of heterogeneous nucleation. Hence, as the concentration of $Zn^{2+}$ existences, a number of larger $Sb_2O_3$.ZnO crystals with aggregated nanosheets composed flower-like morphology form after the reactions [equation (v)]. The shape of calcined $Sb_2O_3$.ZnO MFs is approximately reliable with the growth pattern of antimony doped zinc oxides crystals [38]–[40]. Then the solution was washed thoroughly with acetone, ethanol, and water consecutively and kept for drying at room condition. Finally, the as-grown doped $Sb_2O_3$.ZnO MFs nanomaterials were calcined at 400.0°C for 6 hours in the furnace (Barnstead Thermolyne, 6000 Furnace, USA). The calcined MFs were characterized in detail in terms of their morphological, structural, optical properties, and applied for ethanol chemical sensing based on μ-chips for the first time.

**Fabrication and detection technique of ethanol using $Sb_2O_3$-ZnO MFs on μ-chips**

The sensing area of μ-chip is fabricated with as-grown $Sb_2O_3$ doped ZnO MFs using butyl carbitol acetate (BCA) and ethyl acetate (EA) as a conducting coating agent by drop-coating method. Then it is put into oven at 60.0°C for two hours until the film is completely dry. 0.1 M phosphate buffer solution (PBS) at pH 7.0 is made by mixing 0.2 M $Na_2HPO_4$ and 0.2 M $NaH_2PO_4$ solution in 100.0 mL de-ionize water. A cell is consisted of $Sb_2O_3$-ZnO MFs fabricated μ-chip as a working and Pt connector is used a counter electrodes. As received ethanol is diluted to make various concentrations (0.17 mM to 8.5 M) in PBS solution and used as a target analyte. 70.0 μL of 0.17 mM PBS solution is kept constant during measurements onto the μ-chip. The ratio of current versus concentration (slope of calibration curve) is used to calculate of target ethanol chemical sensitivity. Limit of detection (LOD) is calculated from the ratio of 3N/S versus sensitivity (ratio of noise×3 vs. sensitivity) in the linear dynamic ranges of calibration plot. Electrometer is used as a voltage sources for I–V measurement in simple two electrodes μ-chip system. The as-grown $Sb_2O_3$-ZnO MFs are fabricated and employed for the detection of target analyte.

## Results and Discussion

### *Optical properties of $Sb_2O_3$ doped ZnO MFs*

The optical property of the as-grown $Sb_2O_3$ doped ZnO MFs is one of the significant characteristics for the assessment of its photo-catalytic activity. UV/visible absorption is a technique in which the outer electrons of atoms or molecules absorb radiant energy and undergo transitions to high energy levels. In this phenomenon, the spectrum obtained due to optical absorption can be analyzed to acquire the energy band-gap of the doped metal oxides. For UV/visible spectroscopy, the absorption spectrum of $Sb_2O_3$-ZnO MFs solution is measured as a function of wavelength, which is presented in Figure 1A. It presents a broad absorption band around 289.0 nm in the visible-range between 200.0 to 800.0 nm wavelengths indicating the formation of $Sb_2O_3$-ZnO MFs. Band-gap energy ($E_{bg}$) is calculated on the basis of the maximum absorption band of $Sb_2O_3$-ZnO MFs and found to be ~4.2907 eV, according to following equation (v).

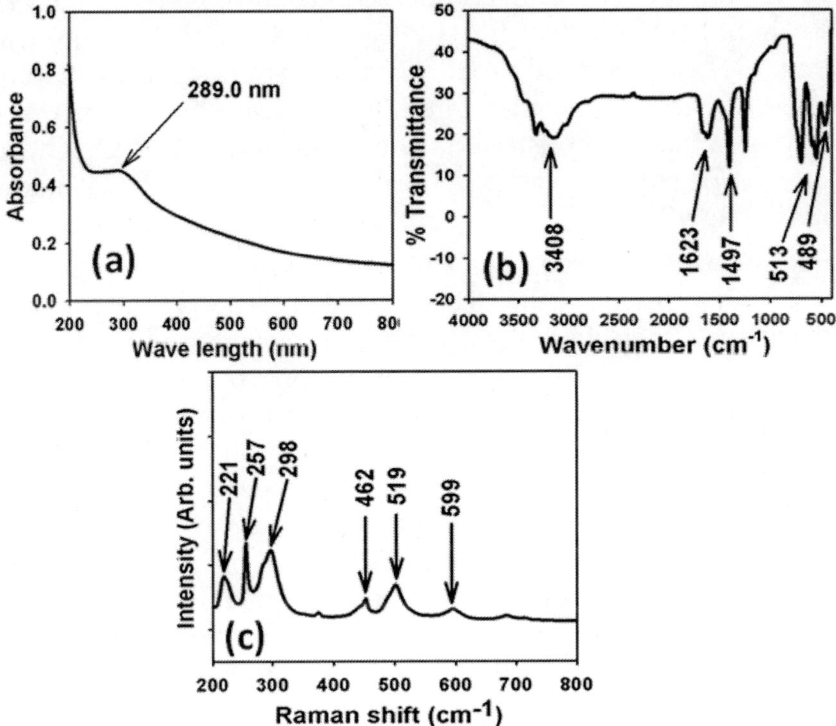


Figure 1. (a) UV/visible, (b) FT-IR, and (c) Raman spectroscopy of as-grown $Sb_2O_3$ doped ZnO MFs at room conditions.

doi:10.1371/journal.pone.0085036.g001

$$Ebg = \frac{1240}{\lambda} \quad (eV) \qquad (v)$$

Where $E_{bg}$ is the band-gap energy and $\lambda_{max}$ is the wavelength (289.0 nm) of the $Sb_2O_3$-ZnO MFs. No extra peak associated with impurities and structural defects are observed in the spectrums, which proved that the synthesized microstructure controlled crystallinity of as-grown $Sb_2O_3$-ZnO MFs [41], [42].

The as-grown $Sb_2O_3$ doped ZnO MFs is also investigated in terms of the atomic and molecular vibrations. To predict the functional-recognition, FT-IR spectra fundamentally in the region of 400~4000

$cm^{-1}$ is investigated at room conditions. Figure 1B displays the FT-IR spectrum of MFs, which represents band at 489, 513, 1497, 1623, and 3408 $cm^{-1}$. These observed broad vibration bands (at 489 & 513 $cm^{-1}$) could be assigned as metal-oxygen (Sb-O & Zn-O mode) stretching vibrations, which demonstrated the configuration of $Sb_2O_3$-ZnO MF materials. The supplementary vibrational bands may be assigned to O-H bending vibration, C-O absorption, and O-H stretching.

The absorption bands at 1497, 1623, and 3408 $cm^{-1}$ generally shows from $CO_2$ and water, which usually semiconductor doped nanostructure materials absorbed from the environment due to their high surface-to-volume ratio of mesoporous nature [43], [44]. Finally, the experimental vibration bands at low frequencies regions recommended the formation of $Sb_2O_3$-ZnO MFs by a facile wet-chemical method. Raman spectroscopy is a spectroscopic technique utilized to display vibrational, rotational, and other low-frequency phases in a Raman active compound. It depends on inelastic scattering of monochromatic light (Raman scattering), usually from a laser in the visible, near infra-red, or near ultra-violet range. The laser light relates with molecular vibrations, phonons or other excitation in the modes, showing in the energy of the laser photons being shifted up or down.

The shift in energy represents the information regarding the phonon modes in the system, where infrared spectroscopy yields similar, but complementary information. Raman spectroscopy is generally established and utilized in material chemistry, since the information is specific to the chemical bonds and symmetry of metal-oxygen stretching or vibrational modes. Usually, there are three vibration modes in $Sb_2O_3$-ZnO MFs nanomaterial crystal: $A_1$, $E_1$ and $E_2$, of which $A_1$ and $E_1$ split into longitudinal ($A_{1L}$, $E_{1L}$) and transverse ($A_{1T}$, $E_{1T}$) ones and $E_2$ contains low and high frequency phonons ($E_{2L}$ and $E_{2H}$) [45], [46]. As-grown $Sb_2O_3$-ZnO MFs is significantly altered the Raman spectra as well as the crystal of ZnO nanostructure [47], [48]. Here, Figure 1c confirms the Raman spectrum, where key aspects of the wave number are employed at about 221, 298, and 257 $cm^{-1}$ for metal-oxygen (Sb-O and Zn-O) stretching vibrations. The large bands can be assigned to a cubic phase of $Sb_2O_3$-ZnO MFs. At 462, 519, and 599 $cm^{-1}$ higher wave-number shifts are revealed owing to the different dimensional effects of the MFs.

Morphological, Structural, and Elemental properties of $Sb_2O_3$ doped ZnO MFs

FE-SEM images of as-grown $Sb_2O_3$ doped ZnO MFs structures are presented in Figure 2(a–d). It exhibits the images of the MFs with micro-dimensional sizes of as-grown $Sb_2O_3$-ZnO MFs. The dimension of MF is calculated in the range of 2.4 μm, which composed of nanosheets (~65.0±10.0 nm). It is clearly exposed from the FE-SEM images that the facile synthesized $Sb_2O_3$-ZnO MFs is microstructures in flower-shape, which is grown in very high-density and possessing almost uniform nanosheet composed MFs. When the size of doped material decreases into micrometer-sized scale, the surface area is increased significantly, this improved the energy of the system and made re-distribution of Zn and Sb ions possible. The micrometer-sized flower could have tightly packed into the lattice, which is an agreement with the publish reports [49], [50]. Crystallinity and crystal phase of $Sb_2O_3$-ZnO MFs were investigated using by powder X-ray diffractometer. Powder X-ray diffraction patterns of doped MFs are represented in Figure 2e. The $Sb_2O_3$.ZnO MFs were investigated and exhibited as face-centered cubic shapes.

Figure 2e reveals characteristic crystallinity of the codoped $Sb_2O_3$.ZnO MFs and their crystalline arrangement, which is investigated using powder X-ray crystallography. All the reflection peaks in this prototype were related with ZnO phase having face-centered cubic zincite geometry [JCPDS # 071-6424]. The phases demonstrated the key features with indices for crystalline ZnO at 2θ values of 32.3(001), 39.8(002), 54.7(020), and 73.2(221) degrees. The face-centered cubic lattice parameters are a = 3.2494, b = 5.2038, and radiation ($CuK_\alpha 1$, λ = 1.5406). The ZnO phases have a high degree of crystallinity. All of the peaks match well with Bragg reflections of the standard zincite structure (point or space-group P63mc) [51]–[53]. The reflected peaks were also found to correspond with $Sb_2O_3$ phase having face-centered cubic orthorhombic geometry [JCPDS # 074-1725]. The phases demonstrated the key features with indices for crystalline $Sb_2O_3$ at 2θ values of 25.7(110), 28.1(111), 32.3(131), 43.6(002), 46.2(242), 59.2(052), 58.3(133), 60.3(072), and 67.4(341) degrees. The $Sb_2O_3$ phases have a high degree of crystallinity. All of the peaks match well with Bragg reflections of the standard orthorhombic structure. These confirmed that there is major number and amount of crystalline codoped $Sb_2O_3$-ZnO present in MFs [54].

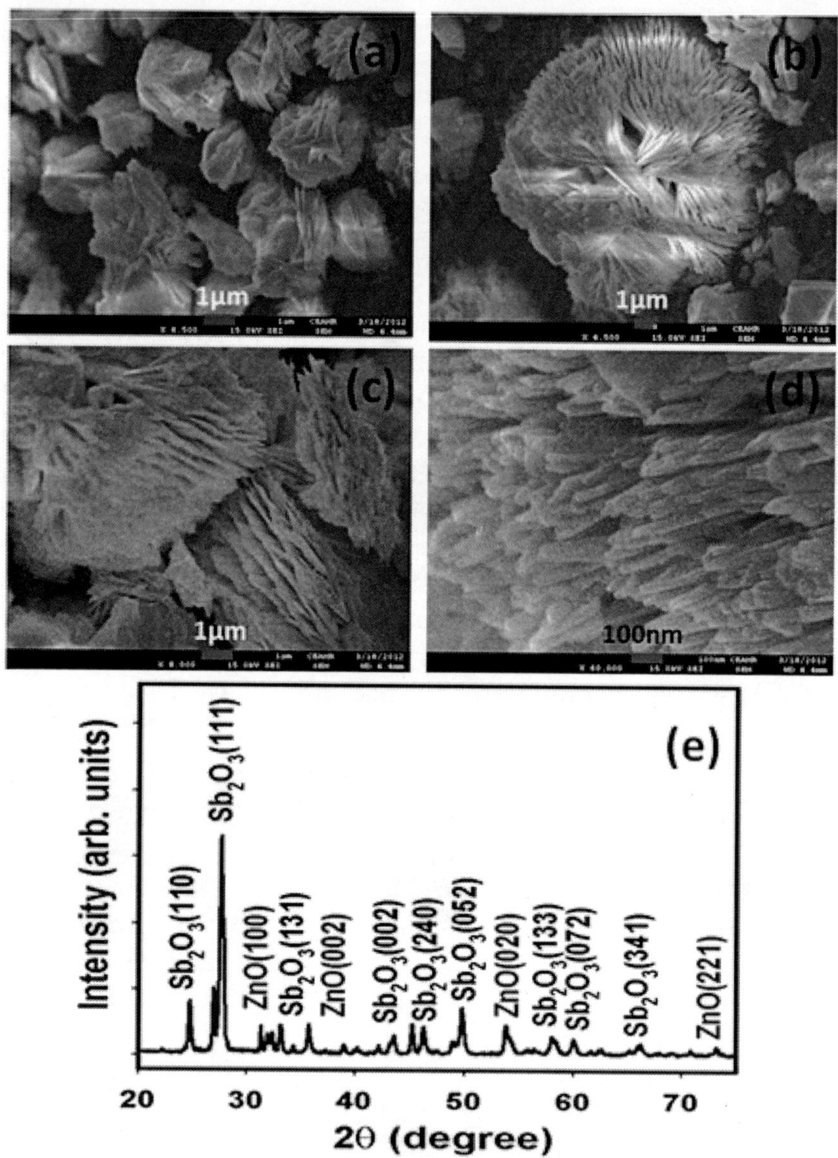


Figure 2. (a–d) FE-SEM images and (e) Powder x-ray diffraction pattern, of as-grown $Sb_2O_3$-ZnO MFs at room conditions.

doi:10.1371/journal.pone.0085036.g002

X-ray photoelectron spectroscopy (XPS) is a quantitative spectroscopic method that determines the chemical-states of the elements that present within doped materials. XPS spectra are acquired by irradiating on a nanomaterial with a beam of X-rays, while simultaneously determining the kinetic energy and number of electrons that get-away from the top one to ten nm of the material being analyzed. Here, XPS measurements were measured for $Sb_2O_3$-ZnO MFs semiconductor nanomaterials to investigate the chemical states of ZnO and $Sb_2O_3$. The XPS spectra of Sb3d, Zn2p, and O1s are presented in Fig. 3a. XPS was also used to resolve the chemical state of the doped $Sb_2O_3$ nanomaterial and their depth. Figure 3b presents the XPS spectra (spin orbit doublet peaks) of the $Sb3d_{(3/2)}$ and $Sb3d_{(1/2)}$ regions recorded with semiconductor doped materials. The binding energy of the $Sb3d_{(3/2)}$ and $Sb3d_{(1/2)}$ peak at 529.1 eV and 539.6 eV respectively denotes the presence of $Sb_2O_3$ since their bindings energies are similar [55]. The O1s spectrum shows a main peak at 531.2 eV in Fig. 3c. The peak at 531.2 eV is assigned to lattice oxygen may be indicated to oxygen (ie, $O_2^-$) presence in the doped $Sb_2O_3$-ZnO MF nanomaterials [56]. In Figure 3d, the spin orbit peaks of the $Zn2p_{(1/2)}$ and $Zn2p_{(3/2)}$ binding energy for all the samples appeared at around 1025 eV and 1048 eV respectively, which is in good agreement with the reference data for ZnO [57].

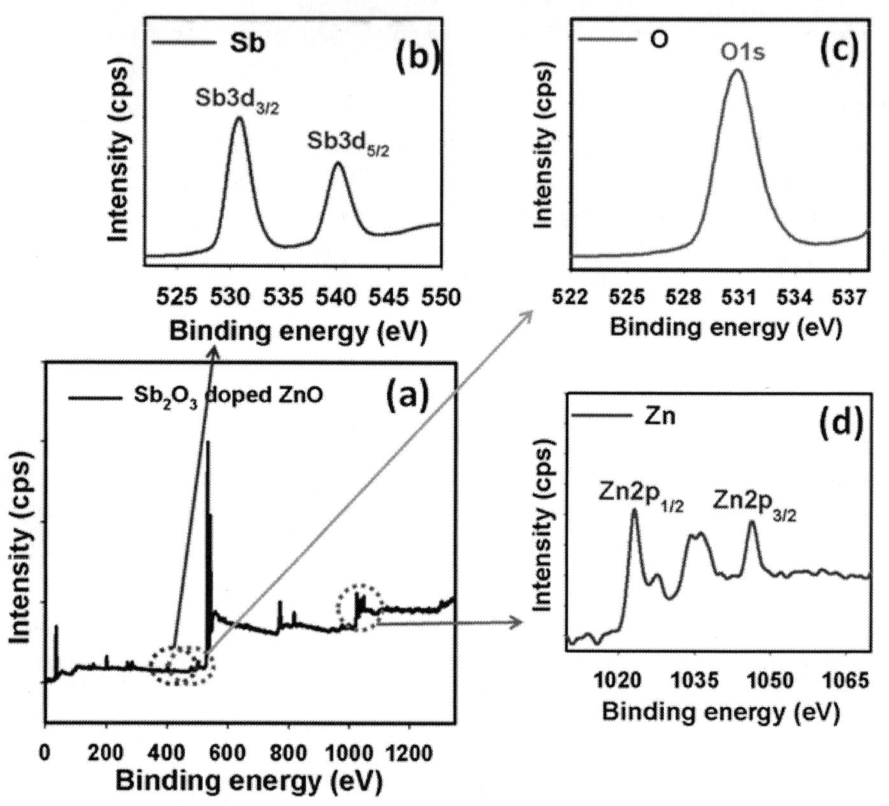


Figure 3. XPS of (a) doped $Sb_2O_3$-ZnO MFs, (b) Sb3d level, (c) O1s level, and (d) Zn2p level acquired with MgKα1 radiation.

doi:10.1371/journal.pone.0085036.g003

Preparation of μ-Chips using photolithography method

Electrochemical μ-chips were fabricated by conventional photolithographic technique, where electrodes and passivation layers are developed on silicon wafer followed by dicing and packaging [58]. Nitrogen-doped silicon wafers are prepared and overflowed by extra-pure water. In this step, all contaminations on the surface and native $SiO_2$ layer are removed perfectly. At first, the wet oxidation is employed and then dry oxidation is executed, where, wafers are annealed in the nitrogen environment. Aluminum is sputtered with aluminum-1% Si target. Then the photolithograph processes are applied. Resist coating, baking, exposure, and development are employed by Kanto chemicals, and then it is rinsed thoroughly by ionic water. Aluminum is etched by etching solution and resistance layer is removed perfectly by plasma etching instrument.

Then silicon wafers are cleaned by acetone, methanol, and finally by plasma simultaneously. Silicon nitride (SiN) layer is deposited by chemical vapor deposition and then pad electrode surfaces are etched by reactive ion etching. Finally residual resist layer is removed by plasma etching. After photolithographic process, platinum is sputtered by SP150-HTS. Then it is patterned by lift-off method, in which wafers are immersed into the remover, and then washed with isopropyl alcohol. Photolithographic process is again investigated, where titanium is sputtered as a binding layer, and then gold is evaporated by deposition method. Finally, gold layer is patterned by lift-off method. Palylene passivation layer is formed for the protection of the μ-chip from water. Photolithographic process is performed again for pad protection. Then palylene-dimer is evaporated by deposition apparatus. Photolithography process is done again for patterning. Palylene layer is patterned by etching. Finally, un-necessary resists are removed by acetone and then wafer is cleaned by isopropyl alcohol (IPA).

Resist is coated on a whole surface of the silicon wafer for protection during dicing process is executed. Silicon wafer is diced into pieces by dicing apparatus and stored into the desiccators, when not in use. Resist on μ-chip surface is removed by acetone and cleaned with isopropyl alcohol (IPA). The opposite side of the chip is roughed by a sandpaper sheet for better adhesion and electrical stability. The μ-chip is bonded with die and packaged by silver paste. It is dried in a drying oven. Pads on chip are connected to the package through gold wire with bonding machine. Finally, silicon-based adhesive is put on the periphery of the chip to protect pads and gold wire from sample solution. Adhesive is dried for 24 hours at room temperature. The semiconductor smart μ-chips were fabricated on silicon wafer. Aluminum was sputtered to fabricate as wiring and bonding pads. Pt-Ti-TiN was sputtered on thermal oxide of silicon and patterned by photolithography to fabricate counter electrode (CE). Ti-TiN layers were used for strong adhesion. Au-Ti were sputtered and lithographed, which made circular working electrode (WE) with a diameter of 1.68 mm in the center of the μ-chip. After electrodes fabrication, palylene layer was fabricated by evaporation method as a passivation layer. The wafer was diced to 5.0 mm square μ-chips. This μ-chip was bonded to a package by silver paste. Aluminum pads were connected to the package by gold wire. Finally, adhesive (Araldite, Hantsman, Japan) was put on the periphery of the chip, which prevents target solution from contacting pads (Figure 4a). The magnified construction view of internal μ-chip center (sensing area) is presented in the Figure 4(b–c).

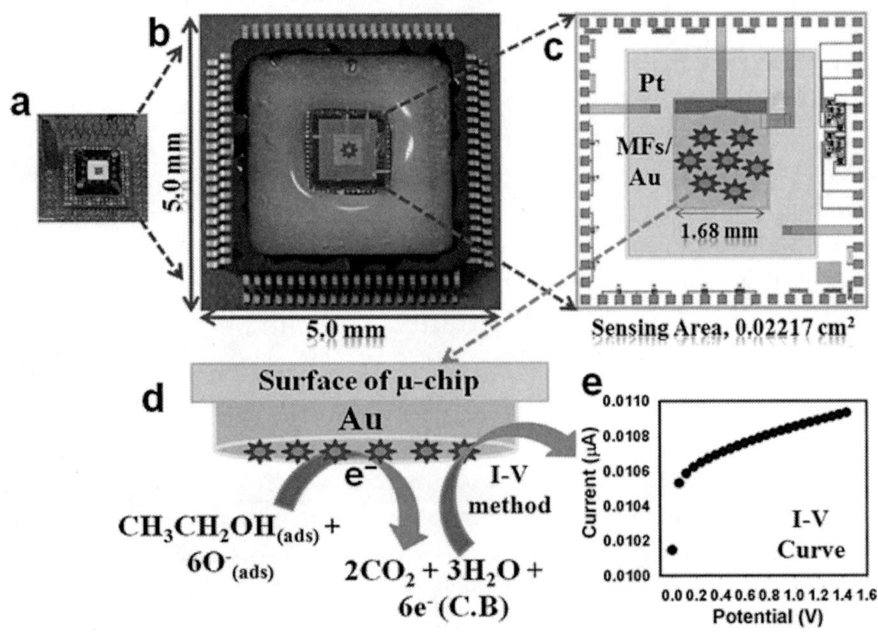


Figure 4. Schematic diagram of (a) real camera-view from top, (b) magnified view of μ-chip, (c) fabrication with $Sb_2O_3$-ZnO MFs with conducting binders (EC & BCA) onto μ-chip sensing-area, (d) reaction mechanism of ethanol in presence of doped $Sb_2O_3$-ZnO MFs, and (e) outcomes of I-V experimental results.

doi:10.1371/journal.pone.0085036.g004

Fabrication and chemical sensor application of $Sb_2O_3$-ZnO/μ-Chips assembly

The potential application of $Sb_2O_3$-ZnO MFs assembled onto μ-chip as chemical sensors (especially ethanol analyte) has been evaluated for measuring and detecting hazardous chemicals, which are not environmental affable. Improvement of doping of these nanosheets composed $Sb_2O_3$-ZnO MFs on μ-chip as chemical sensors is in the initial stage and no other reports are available. The MFs of $Sb_2O_3$-ZnO sensors have advantages such as stability in air, non-toxicity, chemical inertness, electrochemical activity, simplicity to assemble or fabrication, and bio-safe characteristics. As in the case of toxic ethanol sensors, the phenomenon of reason is that the current response in I-V method of $Sb_2O_3$-ZnO MFs considerably changes when aqueous ethanol are adsorbed. The calcined $Sb_2O_3$-ZnO MFs were applied for modification of chemical sensor, where ethanol was measured as target analyte. The fabricated-surface of $Sb_2O_3$-ZnO MFs sensor was made with conducting binders (EC & BCA) on the μ-chip surface, which is presented in the Figure 4(c–d). The fabricated μ-chip electrode was placed into the oven at low temperature (50.0°C) for 2 hours to make it dry, stable, and uniform the surface totally. I–V signals of chemical sensor are anticipated having $Sb_2O_3$-ZnO doped thin film as a function of current versus potential for hazardous ethanol. The real electrical responses of target ethanol are investigated by simple and reliable I–V technique using $Sb_2O_3$-ZnO MFs fabricated μ-chip, which is presented in Figure 4e.The time holding of electrometer was set for 1.0 sec. A significant amplification in the current response with applied potential is noticeably confirmed. The simple, reliable, possible reaction mechanism is generalized in Scheme 1d in presence of ethanol on $Sb_2O_3$-ZnO MFs sensor surfaces by I–V technique. The ethanol is converted to water and carbon dioxide in presence of doped nanomaterials by releasing electrons ($-6e^-$) to the reaction system (conduction band, C.B.), which improved and enhanced the current responses against potential during the I–V measurement at room conditions.

Figure 5a shows the current responses of un-coated (gray-dotted) and coated (dark-dotted) μ-chip working electrodes with $Sb_2O_3$-ZnO MFs in absence of target ethanol. With nanosheets composed MFs fabricating surface, the current signal is slightly reduced compared to uncoated μ-chip surface, which indicates the surface is slightly blocked with doped MF nanomaterials in the buffer system. The current changes for the un-coated μ-chips (dark-dotted) towards target ethanol (~50.0 μL), and MF nanomaterials modified film before (deep-blue-dotted) and after (light-blue-dotted) injecting of target

50.0 μL ethanol (~0.17 mM) onto $Sb_2O_3$-ZnO MFs modified μ-chips is showed in Figure 5b. A significant current enhancement is exhibited with the $Sb_2O_3$-ZnO MFs modified μ-chips compared with uncoated μ-chips due to the presence of micro-structures, which has higher-specific surface area, larger-surface coverage, excellent absorption and adsorption capability into the porous MF surfaces towards the target ethanol.

This significant change of surface current is examined in every injection of the target ethanol onto the doped modified μ-chips by electrometer. I–V responses with doped $Sb_2O_3$-ZnO MFs modified μ-chip surface are investigated from the various concentrations (0.17 mM to 8.5 M) of ethanol, which is showed in Figure 5c. It shows the current changes of fabricated μ-chip films as a function of ethanol concentration in room condition. It was also found that at low to high concentration of target analyte, the current responses were enhanced regularly. The potential current changes at lower to higher potential range (potential, +0.10 V to +1.3 V) based on various analyte concentration are observed, which is clearly presented in Figure 5c. A large range of analyte concentration is measured the probable analytical limit, which is calculated in 0.17 mM to 8.5 M. The calibration (at +0.3V) and magnified-calibration curves are plotted from the various ethanol concentrations, which are presented in the Figure 5(d–e). The sensitivity is estimated from the calibration curve, which is close to ~5.848 $\mu Acm^{-2}mM^{-1}$. The linear dynamic range of this sensor displays from 0.17 mM to 0.85 M (linearity, R = 0.9989) and the detection limit was considered as 0.11±0.001 mM [3×noise (N)/slope(S)].

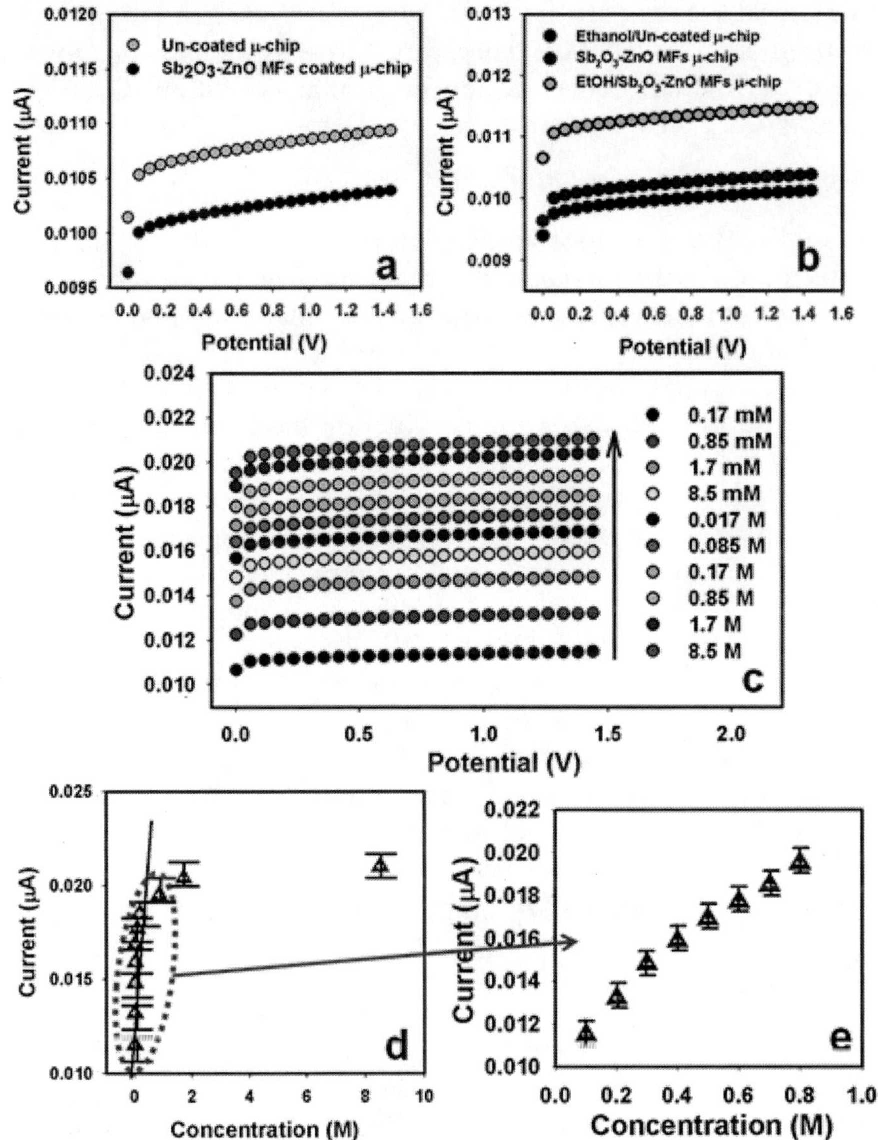


**Figure 5**. I–V responses of (a) un-coated and $Sb_2O_3$-ZnO MFs coated μ-chip without ethanol; (b) with 0.17 mM ethanol for un-coated μ-chip; without ethanol for $Sb_2O_3$-ZnO MFs coated μ-chip; and with 0.17 mM ethanol for $Sb_2O_3$-ZnO MFs coated μ-chip; (c) concentration variations (0.17 mM to 8.5 M) of analyte; and (d) calibration plot of doped nanomaterial fabricated on μ-chip surfaces.

Potential was chosen in 0.1 to +1.4 V ranges. Error limit of I-V measurement was ± 0.001. There are three trial has been done in same experimental concentration at similar condition. Coefficient variation (CV): 0.1699.

doi:10.1371/journal.pone.0085036.g005

Usually, the resistance value of doped semiconductor materials are decreased with increasing surrounding active oxygen, which is the fundamental characteristics of nanomaterials [59]. Actually, oxygen adsorption demonstrates an significant responsibility in the electrical properties of the $Sb_2O_3$-ZnO MFs onto μ-chip. The oxygen ion adsorption is removed the conduction electrons and increased the resistance of $Sb_2O_3$-ZnO MFs. Unstable oxygen species (i.e., $O_2^-$ & $O^-$) are adsorbed on the doped MF surface at room temperature, and the quantity of such chemisorbed oxygen species is directly depended on morphological and structural properties. At room condition, $O_2^-$ is chemisorbed, while on nanosheets composed microflowers morphology, $O_2^-$ and $O^-$ are chemisorbed significantly. For this reason, the active $O_2^-$ is disappeared quickly [60]. Here, ethanol sensing mechanism on $Sb_2O_3$-ZnO MFs/μ-chip sensor is executed due to the presence of semiconductors oxides. The oxidation or reduction of the semiconductor MFs is held, according to the dissolved $O_2$ in bulk-solution or surface-air of the neighboring atmosphere according to the following equations (vi–viii).

$\mathbf{O_{2(diss)}}$ ($Sb_2O_3$–ZnO MFs/μ-chip) → $\mathbf{O_{2(ads)}}$ **(vi)**

$\mathbf{e^-}$ ($Sb_2O_3$–ZnO MFs/μ-chip)+$\mathbf{O_2}$ → $\mathbf{O_2^-}$ **(vii)**

$\mathbf{e^-}$ ($Sb_2O_3$–ZnO MFs/μ-chip)+$\mathbf{O_2^-}$ → $\mathbf{2O^-}$ **(viii)**

These reactions are held in bulk-system or air/liquid interface or adjacent atmosphere due to the small carrier concentration which enhanced the resistances. The ethanol sensitivity could be attributed to the high oxygen deficiency on $Sb_2O_3$-ZnO MFs/μ-chip (eg. $MO_x$) and higher density conducts to increase oxygen adsorption. Larger the quantity of oxygen adsorbed on the fabricated sensor surface, larger would be the oxidizing potential as well as faster would be the oxidation of ethanol. The reactivity of ethanol would have been very large as compared to other fabricated material surfaces surface under identical condition [61]–[63]. When ethanol reacts with the adsorbed oxygen on the exterior/interior of the $Sb_2O_3$-ZnO MFs/μ-chip layer, it oxidized to carbon dioxide and water by releasing free electrons ($6e^-$) in the conduction band, which is expressed through the following reactions (ix).

$$CH_3CH_2OH_{(ads)}+6O^-_{(ads)} \rightarrow 2CO_2+3H_2O+6e^-(C.B.) \quad (ix)$$

In the reaction system, these reactions referred to oxidation of the reducing carriers. This method is enhanced the carrier concentration and consequently decreased the resistance on adjacent reducing analytes. The elimination of ionosorbed oxygen amplified the electron concentration onto $Sb_2O_3$-ZnO MFs/μ-chip and hence the surface conductance is increased in the film [64], [65]. The reducing analyte (ethanol) gives electrons to $Sb_2O_3$-ZnO MFs/μ-chip surface. Consequently, resistance is reduced, and hence the conductance is increased. This is the cause why the analyte response (current) amplifies with increasing potential. Thus produced electrons contribute to rapid increase in conductance of the thick $Sb_2O_3$-ZnO MFs/μ-chip film. The $Sb_2O_3$-ZnO MFs unusual regions dispersed on the surface would progress the capability of nanomaterial to absorb more oxygen species giving high resistance in air ambient, which is presented in Figure 6.

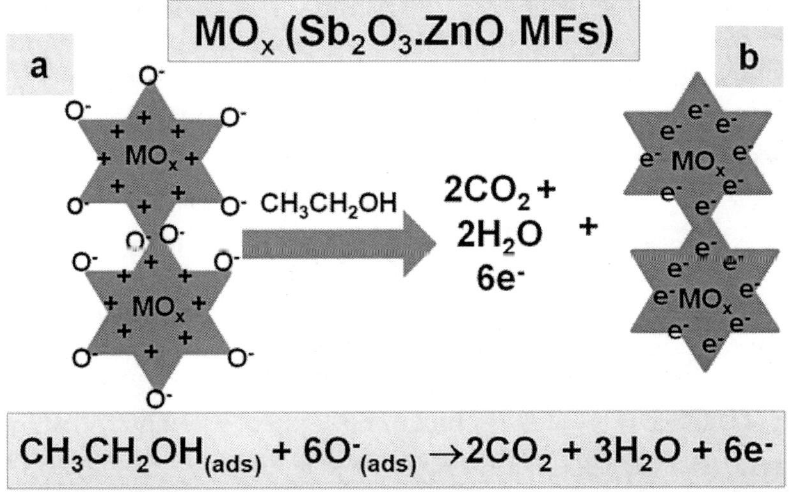


Figure 6. Mechanism of $Sb_2O_3$-ZnO MFs/μ-chip ethanol chemical sensors at ambient conditions.

doi:10.1371/journal.pone.0085036.g006

On the other approach, the utmost ethanol response of $Sb_2O_3$-ZnO MFs/μ-chip was attributed to the larger chemical communication on the sensing surface due to the larger surface area and meso-porous natures. The high ethanol response of the $Sb_2O_3$-ZnO MFs can be understandable in more detail in relative to the probable chemical-sensing mechanism, in terms of p-type doped semiconductor nano-materials. The oxide surface of an p-type semiconductor is readily covered with chemisorbed oxygen [66], [67]. Therefore, at identical condition, the adsorption of negatively charged oxygen can generate the holes for conduction. The subsequent ethanol-sensing reactions might be considered according to the charges of the adsorbed oxygen species ($Sb_2O_3$-ZnO MFs/μ-chip) under the statement of full oxidation of $C_2H_5OH$ according to the following equations (x-xi).

$$\mathbf{{}^1\!/_2 O_{2(g/l)} \leftrightarrow O^-_{(ads)}}\ (Sb_2O_3\text{–}ZnO\ MFs/\mu\text{–chip})\mathbf{+h^\circ}\ \mathbf{(x)}$$

$$\mathbf{C_2H_5OH_{(g)}+6O^-_{(ads)}+6h^\circ \rightarrow 2CO_{2(g)}+3H_2O_{(g)}\ (xi)}$$

The oxidation reaction with reducing ethanol amplifies the resistivity of the surface regions of the p-type doped $Sb_2O_3$-ZnO MFs/μ-chip, which in turn enhances the sensor resistance. The resistive contacts among the $Sb_2O_3$-ZnO MFs nano-materials control the chemi-sensor resistance. Therefore, the ethanol response is extensively dependent upon the dimensions of the nanosheet composed MFs, the large-active surface area and the nano-porosity. According to the charge accumulation reproduction of p-type semiconductors, the conduction occurs along the conductive as well as active sensor surface of $Sb_2O_3$-ZnO MFs/μ-chip [68]–[70].

The sensor response time was ~10.0 sec for the $Sb_2O_3$-ZnO MFs coated μ-chip sensor to achieve saturated steady state current in I–V plots. The major sensitivity of μ-chip sensor can be attributed to the good absorption (porous surfaces MFs fabricated with binders), adsorption ability, high-catalytic activity, and good bio-compatibility of the $Sb_2O_3$-ZnO MFs/μ-chip. The expected sensitivity of the MF fabricated sensor is relatively better than previously reported ethanol sensors based on other composites or materials modified electrodes [71]. Due to perceptive surface area, here the doped nano-materials proposed a beneficial microenvironment for the toxic chemical detection (by adsorption) and recognition with excellent quantity. The prominent sensitivity of $Sb_2O_3$-ZnO MFs affords high electron communication features which improved the direct electron communication between the active sites of nano-sheets composed microstructures and μ-chips. The modified thin $Sb_2O_3$-ZnO MFs/μ-chip sensor film had a better reliability as well as stability in ambient conditions. $Sb_2O_3$-ZnO MFs/μ-chip exhibits several approaching in providing ethanol chemical based sensors, and encouraging improvement has been accomplished in the research section.

To check the reproducibly and storage stabilities, I–V response for $Sb_2O_3$-ZnO MFs coated μ-chip sensor was examined (up to 2 weeks). After each experiment, the fabricated $Sb_2O_3$-ZnO MFs/μ-chip substrate was washed thoroughly with the PBS buffer solution and observed that the current response was not significantly decreased (Figure 7a). The sensitivity was retained almost same of initial sensitivity up to week ($1^{st}$ to $2^{nd}$ week), after that the response of the fabricated electrode gradually decreased. A series of six successive measurements of 0.17 mM ethanol in 0.1 mM PBS yielded a good reproducible signal at $Sb_2O_3$-ZnO MFs/μ-chip sensor in different conditions with a relative standard deviation (RSD) of 3.7% (Figure not shown). The sensor-to-sensor and run-to-run repeatability for 0.17 mM ethanol detection were found to be 1.9% using $Sb_2O_3$-ZnO MFs/μ-chip. To investigate the long-term storage stabilities, the response for the MFs sensor was determined with the respect to the storing time. The long-term storing stability of the $Sb_2O_3$-ZnO MFs/μ-chip sensor was investigated significantly at room conditions. The sensitivity retained 94% of initial sensitivity for several days. The above results clearly suggested that the fabricated sensor can be used for several weeks without any significant loss in sensitivity. The dynamic response (0.17 mM to 0.85 M) of the sensor was investigated from the practical concentration variation curve. The sensor response time is mentioned and investigated using this sensor system at room conditions. In this study, it shows the MFs/Chips-based ethanol sensor response after successive injection in buffer solutions containing 1.0 mM of ethanol, which is presented in Figure 7a. A recovery time is observed with a small loss of signal (~0.007uA), suggesting that the sensor can be re-used further [Figure 7b (inset)]. It was also investigated the sensing selectivity performances (interferences) with other chemicals like methanol, acetone, chloroform, dichloromethane, phenyl hydrazine, nitrophenol etc. Ethanol exhibited the maximum current response by I–V system using MFs fabricated micro-chip electrode compared to methanol and others. It was specific towards ethanol compared to all other chemicals. A comparative study with all chemicals using $Sb_2O_3$-ZnO MFs/μ-chip is included in Figure 7c. In Table 1, it is compared the performances for ethanol chemical detection based $Sb_2O_3$-ZnO MFs/μ-chip using various modified electrode materials [72]–[80].

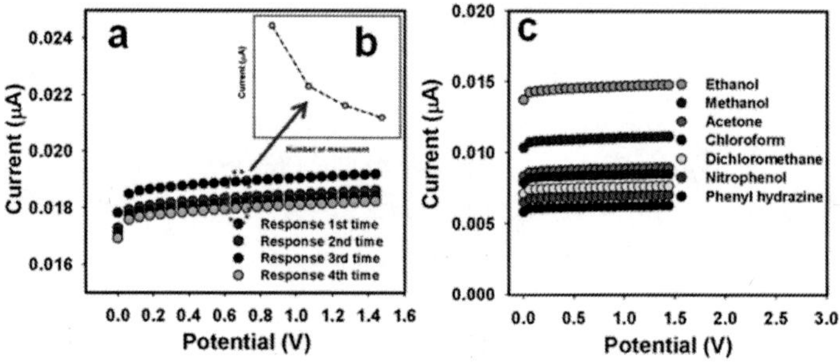


Figure 7. I–V responses of $Sb_2O_3$-ZnO MFs coated μ-chip are presented for ethanol sensors reproducibility (a), sensor-responses (inset, b), and selectivity (c) study.

Ethanol and other chemicals concentration are taken as1.0 mM for selectivity study.

doi:10.1371/journal.pone.0085036.g007

Table 1. Comparison the performances for ethanol detection based on various nanomaterial fabricated electrodes.

doi:10.1371/journal.pone.0085036.t001

| Materials | Methods | Linear Dynamic Range, LDR | Sensitivity | Linearity, $r^2$ | Limit of Detection, LOD | Response Time | References |
|---|---|---|---|---|---|---|---|
| Ni/Pt/Ti | Potential amperometry | – | 3.08 μAmM$^{-1}$ cm$^{-2}$ | – | – | – | [72] |
| Ni-doped $SnO_2$ nanostructure | I-V | 1.0 nM–1.0 mM | 2.3148 μA cm$^{-2}$mM$^{-1}$ | 0.8440 | 0.6 nM | 10.0 s | [73] |
| Pd–Ni/SiNWs electrode | Potential amperometry | – | 0.76 mAmM$^{-1}$cm$^{-2}$ | 0.9970 | 10.0 μM | – | [74] |
| ZnO-$CeO_2$ Nanoparticles | I-V | 1.7 mM–1.7 M | 2.1949 μA cm$^{-2}$mM$^{-1}$ | 0.9463 | 0.6±0.05 mM | 10.0 s | [75] |
| RuO-modified Ni electrode | Cyclic voltammetry | 100–1000 ppm | 4.92 μAppm$^{-1}$cm$^{-2}$ | – | – | 13.0 s | [76] |
| $CeO_2$ nanoparticles | I-V | 0.17 mM–0.17 M | 0.92 μAcm$^{-2}$mM$^{-1}$ | 0.7458 | 0.124±0.010 mM | 10.0 s | [77] |
| Al-doped ZnO nanomaterial | I-V method | Up to 3000 ppm | 1000 ppm ethanol | – | – | ~8.0 s | [78] |
| CuO nanosheets | I-V | up to 1.7 M | ~0.9722 μAcm$^{-2}$mM$^{-1}$ | 0.7806 | 0.143 mM | 10.0 s | [79] |
| Sm-Doped $Co_3O_4$ Nanokernels | I-V | 1.0 nM–10.0 mM | 2.1991±0.10 μAcm$^{-2}$mM$^{-1}$ | 0.9065 | 0.63±0.02 nM | 10.0 s | [80] |
| **$Sb_2O_3$-ZnO MFs** | **I-V** | **0.17 mM–0.85 M** | **5.845 uA cm$^{-2}$mM$^{-1}$** | **0.9989** | **0.11±0.02 mM** | **10.0 s** | **Current work** |

doi:10.1371/journal.pone.0085036.t001

# CONCLUSION

Transition-metal doped semiconductor $Sb_2O_3$-ZnO MFs are prepared by easy, simple, efficient, reliable, and economical approaches using reducing agents. The structural, morphological, and optical properties are performed by using XRD, XPS, FE-SEM, and UV-visible techniques respectively. The $Sb_2O_3$-ZnO MFs/μ-chip has assembled by simple fabricated method and displayed higher sensitivity for chemical sensing. They are efficiently prepared for sensitive ethanol sensor based on $Sb_2O_3$-ZnO MFs embedded μ-chips with conducting coating binders, for the first time. The analytical performances of the fabricated ethanol MFs sensors are excellent in terms of sensitivity, detection limit, linear dynamic ranges, and in short response time. $Sb_2O_3$-ZnO MFs/μ-chips are exhibited higher-sensitivity (~5.845uA $cm^{-2}mM^{-1}$) and lower-detection limit (~0.11±0.02 mM) with good linearity in short response time, which efficiently utilized as chemi-sensor for ethanol onto μ-chips. This novel attempt is introduced a well-organized route of efficient chemical sensor development for environmental toxic pollutants and health-care fields in broad scale.

## Acknowledgments

This work was funded by the Deanship of Scientific Research (DSR), King Abdulaziz University, Jeddah, under grant No. **130-019-D1433**. The authors, therefore, acknowledge with thanks DSR technical and financial support.

## Author Contributions

Conceived and designed the experiments: MMR. Performed the experiments: MMR SBK. Analyzed the data: MMR SBK AMA. Contributed reagents/materials/analysis tools: MMR SBK AMA. Wrote the paper: MMR.

# REFERENCES

1. Kumar A, Singhal A (2007) Synthesis of colloidal β-$Fe_2O_3$ nanostructures-influence of addition of $Co^{2+}$ on their morphology and magnetic behavior. Nanotechnol 18: 475703–475710. doi: 10.1088/0957-4484/18/47/475703

2. Whitesides GM, Boncheva M (2002) Beyond molecules: Self-assembly of mesoscopic and macroscopic components. Proc Natl Acad Sci USA 99: 4769–4774. doi: 10.1073/pnas.082065899

3. Dale L (2009) Huber (2009) Synthesis, properties, and applications of iron NPs. Small 1: 482–501. doi: 10.1002/smll.200500006

4. Shaalan NM, Yamazaki T, Kikuta T (2011) Influence of morphology and structure geometry on $NO_2$ gas-sensing characteristics of $SnO_2$ nanostructures synthesized via a thermal evaporation method. Sens Actuator B 153: 11–16. doi: 10.1016/j.snb.2010.09.070

5. Rahman MM, Jamal A, Khan SB, Faisal M, Asiri AM (2012) Highly Sensitive methanol chemical sensor based on undoped silver oxide nanoparticles prepared by a solution method. Microchim Acta 178: 99–106. doi: 10.1007/s00604-012-0817-2

6. Umar A, Rahman MM, Kim SH, Hahn YB (2008) Zinc oxide nanonail based chemical sensor for hydrazine detection. Chem Commun 166–169.

7. Rahman MM, Jamal A, Khan SB, Faisal M (2011) Characterization and applications of as-grown β-$Fe_2O_3$ nanoparticles prepared by hydrothermal method. J Nanopart Res 13: 3789–3799. doi: 10.1007/s11051-011-0301-7

8. Yao C, Zeng Q, Goya GF, Torres T, Liu J, et al. (2007) $ZnFe_2O_4$ nanocrystals: synthesis and magnetic properties. J Phys Chem C 111: 12274–12278. doi: 10.1021/jp0732763

9. Ingler-Jr WB, Baltrus JP, Khan SUM (2004) Photoresponse of p-Type Zinc-Doped Iron(III) Oxide Thin Films. J Am Chem Soc 126: 10238–10239. doi: 10.1021/ja048461y

10. Holmes JD, Johnston KP, Doty RC, Korgel BA (2000) Control of Thickness and Orientation of Solution-Grown Silicon Nanowires. Science 287: 1471–1473. doi: 10.1126/science.287.5457.1471

11. Pan Z, Dai Z, Wang ZL (2001) Nanobelts of semiconducting oxides. Science 291: 1947–1949. doi: 10.1126/science.1058120

12. Wang N, Tang Y, Zhang Y, Lee CS, Lee ST (1998) Nucleation and growth of Si nanowires from silicon oxide. Phys. Rev. B 58: R16024–R16026. doi: 10.1103/physrevb.58.r16024

13. Gui L, Wu Z, Liu T, Wang W, Zhu H (2000) Synthesis of novel $Sb_2O_3$ and $Sb_2O_5$nanorods. Chem. Phys. Lett 318: 49. doi: 10.4028/www.scientific.net/kem.317-318.849

14. Friedrichs S, Meyer RR, Sloan J, Kirkland AI, Hutchison JL, et al. (2001) Complete characterisation of a $Sb_2O_3$/(21,−8)SWNT inclusion composite. Chem. Commun. 929–930.

15. Suchea M, Christoulakis S, Katsarakis N, Kitsopoulos T, Kiriakidis G (2007) Comparative study of zinc oxide and aluminium doped zinc oxide transparent thin films by direct current magnetron sputtering. Thin Sol Film 515: 6562. doi: 10.1016/j.tsf.2006.11.151

16. Lee SR, Rahman MM, Ishida M, Sawada K (2009) Development of highly sensitive acetylcholine sensor based on acetylcholine by charge transfer

techniques esterase using smart biochips. Trends Anal Chem 28: 196. doi: 10.1016/j.trac.2008.11.009

17. Pearton SJ, Heo WH, Ivill M, Norton DP, Steiner T (2004) Dilute magnetic semiconducting oxides. Semicond Sci Technol 19: 59. doi: 10.1088/0268-1242/19/10/r01
18. Nomura K, Ohta H, Ueda K, Kamiya T, Hirano M, et al. (2003) Thin-film transistor fabricated in single-crystalline transparent oxide semiconductor. Science 300: 1269. doi: 10.1126/science.1083212
19. Umar A, Rahman MM, Vaseem M, Hahn YB (2009) Ultra-Sensitive Cholestrol Biosensor Based on Low Temperature Grown ZnO Nanoparticles. Electrochem. Commun 11: 118. doi: 10.1016/j.elecom.2008.10.046
20. Rahman MM, Jamal A, Khan SB, Faisal M (2011) CuO Codoped ZnO Based Nanostructured Materials for Sensitive Chemical Sensor Applications. ACS Appl Mater Interfac 3: 1346. doi: 10.1021/am200151f
21. Wang ZLJ (2004) Zinc oxide nanostructures: growth, properties and applications. Phys Condens Matt 16: 829.
22. Ng HT, Han J, Yamada T, Nguyen P, Chen YP, et al. (2004) Single Crystal Nanowire Vertical Surround-Gate Field-Effect Transistor. Nano Lett 4: 1247. doi: 10.1021/nl049461z
23. Soci C, Zhang A, Xiang B, Dayeh SA, Aplin DPR, et al. (2007) ZnO Nanowire UV Photodetectors with High Internal Gain. Nano Lett 7: 1003. doi: 10.1021/nl070111x
24. Lee SR, Rahman MM, Ishida M, Sawada K (2009) Fabrication of a Highly Sensitive Penicillin Sensor Based on Charge Transfer Techniques. Biosens Bioelectron 24: 1877. doi: 10.1016/j.bios.2008.09.008
25. Huang MH, Mao S, Feick H, Yan H, Wu Y, et al. (2001) Room temperature ultraviolet nanowire nanolasers. Science 292: 1897. doi: 10.1126/science.1060367
26. Wang XD, Song JH, Liu J, Wang ZL (2007) Direct-Current Nano-generator Driven by Ultrasonic Waves. Science 316: 102. doi: 10.1126/science.1139366
27. Lee SR, Rahman MM, Ishida M, Sawada K (2009) Development of highly sensitive acetylcholine sensor based on acetylcholine by charge transfer techniques esterase using smart biochips. Trends in Anal. Chem 28: 196–203. doi: 10.1016/j.trac.2008.11.009
28. Khan SB, Rahman MM, Jang ES, Akhtar K, Han H (2011) Special susceptive aqueous ammonia chemi-sensor: Extended applications of novel UV-curable polyurethane-clay nanohybrid. Talanta 84: 1005–1010. doi: 10.1016/j.talanta.2011.02.036
29. Yang Z, Huang Y, Chen G, Guo Z, Cheng S, et al. (2009) Ethanol gas sensor based on Al-doped ZnO nanomaterial withmany gas diffusing channels. Sens Actuator B Chem 140: 549–556. doi: 10.1016/j.snb.2009.04.052
30. Gonga H, Hu JQ, Wang JH, Ong CH, Zhu FR (2006) Nano-crystalline Cu-doped ZnO thin film gas sensor for CO. Sens Actuator B: Chem 115: 247–251. doi: 10.1016/j.snb.2005.09.008

31. Francesco FD, Fuoco R, Trivella MG, Ceccarini A (2005) Breath analysis:trends in techniques and clinical applications. Microchem J 79: 405. doi: 10.1016/j.microc.2004.10.008

32. GW, Park KY, Anwar MS, Seo YJ, Sung CH, et al. (2012) Preparation and properties of $Sb_2O_3$-doped $SnO_2$ thin films deposited by using PLD. J Kor Phys Soc 60: 1548–1551.

33. Mori H, Sakata H (1996) Oxygen gas-sensing properties of $V_2O_5$-$Sb_2O_3$-$TeO_2$ glass. Materials Chemistry and Physics 45: 211–215. doi: 10.4028/www.scientific.net/amr.211-212.152

34. Montenegro A, Ponce MA, Castro MS, Rodríguez-Paez JEJ (2007) $SnO_2$-$Bi_2O_3$ and $SnO_2$-$Sb_2O_3$ gas sensors obtained by soft chemical method. J European Ceram Soc 27: 4143–4146. doi: 10.3989/cyv.2008.v47.i1.208

35. Rahman MM, Umar A, Sawada K (2009) Development of Amperometric Glucose Biosensor Based on Glucose Oxidase Enzyme Immobilized with Multi-Walled Carbon Nanotubes at Low Potential. Sens Actuator B 137: 327–333. doi: 10.1016/j.snb.2008.10.060

36. Nicoletti S, Zampolli S, Elmi I, Dori L, Severi M (2003) Use of different sensing materials and deposition techniques for thin-film sensors to increase sensitivity and selectivity. IEEE Sens J 3: 454–459. doi: 10.1109/jsen.2003.815797

37. Aguilar-Leyva A, Maldonado, De-la-Olvera M (2007) Gas-sensing characteristics of undoped-SnO2 thin films and Ag/SnO2 and SnO2/Ag structures in a propane atmosphere. Mater Character 58: 740–744. doi: 10.1016/j.matchar.2006.11.016

38. Zhang C, Hu Y, Lu W, Cao M, Zhou D (2002) Influence of $TiO_2$/$Sb_2O_3$ ratio on ZnO varistor ceramics. J Europ Ceram Soc 22: 61–65. doi: 10.1016/s0955-2219(01)00232-1

39. Das S, Kim DY, Choi CM, Hahn YB (2011) Influence of aqueous hexamethylenetetramine on the morphology of self-assembled $SnO_2$ nanocrystals. Mater Res Bull 46: 609–614. doi: 10.1016/j.materresbull.2010.12.034

40. Lao YW, Kuo ST, Tuan WH (2007) Effect of $Bi_2O_3$ and Sb2O3 on the grain size distribution of ZnO. J. Electroceram 19: 187–194. doi: 10.1007/s10832-007-9187-2

41. Vicente FSD, Li MS, Nalin M, Messaddeq Y (2003) Photoinduced structural changes in antimony polyphosphate based glasses. J. Non-Crystall. Sol 330: 168–173. doi: 10.1016/j.jnoncrysol.2003.08.047

42. Han N, Chai L, Wang Q, Tian Y, Deng P, et al. (2010) Evaluating the doping effect of Fe, Ti and Sn on gas sensing property of ZnO. Sens Actuator B Chem 147: 525–530. doi: 10.1016/j.snb.2010.03.082

43. Gibert JP, Cuesta JML, Bergeret A, Crespy A (2000) Study of the degradation of fire-retarded PP/PE copolymers using DTA/TGA coupled with FTIR. Poly Degrad Stability 67: 437–447. doi: 10.1016/s0141-3910(99)00142-1

44. Li Y, Yi R, Yan A, Deng L, Zhou K, et al. (2009) Facile synthesis and properties

of $ZnFe_2O_4$ and $ZnFe_2O_4$/polypyrrole core-shell nanoparticles. Sol Stat Sci 11: 1319–1324. doi: 10.1016/j.solidstatesciences.2009.04.014

45. Damen TC, Porto SPS, Tell B (1966) Raman-effect in zinc oxide. Phys Rev 142: 570–574. doi: 10.1103/physrev.142.570

46. Rahman MM, Khan SB, Faisal M, Asiri AM, Tariq M A (2012) Detection of Aprepitant Drug Based on Low-dimensional Un-doped Iron oxide Nanoparticles Prepared by Solution method. Electrochim. Acta 75: 164–170. doi: 10.1016/j.electacta.2012.04.093

47. Bundesmann C, Ashkenov N, Schubert M, Spemann D, Butz T, et al. (2003) Raman scattering in ZnO thin films doped with Fe, Sb, Al, Ga, and Li. Appl Phys Lett 83: 1974–1976. doi: 10.1063/1.1609251

48. Cebriano T, Méndez B, Piqueras J (2012) Micro- and nanostructures of $Sb_2O_3$ grown by evaporationedeposition: Self assembly phenomena, fractal and dendritic growth. Mater Chem Phys 135: 1096–1103. doi: 10.1016/j.matchemphys.2012.06.024

49. Rahman MM, Jamal A, Khan SB, Faisal M (2011) Characterization and Applications of as-grown $\beta$-$Fe_2O_3$ Nanoparticles Prepared by Hydrothermal Method. J Nanopart Res 13: 3789–3799. doi: 10.1007/s11051-011-0301-7

50. Rahman MM, Khan SB, Faisal M, Asiri AM, Alamry KA (2012) Highly Sensitive Formaldehyde Chemical Sensor Based on Hydrothermally Prepared Spinel $ZnFe_2O_4$Nanorods. Sens Actuator B: Chem 171–172: 932–937. doi: 10.1016/j.snb.2012.06.006

51. Umar A, Rahman MM, Kim SH, Hahn YB (2008) ZnO Nanonails: Synthesis and Their Application as Glucose Biosensor. J Nanosci Nanotech 8: 3216. doi: 10.1166/jnn.2008.116

52. Faisal M, Khan SB, Rahman MM, Jamal A, Asiri AM, Abdullah MM (2012) Fabrication of ZnO nanoparticles based sensitive methanol sensor and efficient photocatalyst. App Surf Sci 258: 7515–7522. doi: 10.1016/j.apsusc.2012.04.075

53. Khan SB, Faisal M, Rahman MM, Jamal A (2011) Low-temperature Growth of ZnO Nanoparticles: Photocatalyst and Acetone Sensors. Talanta 85: 943–949. doi: 10.1016/j.talanta.2011.05.003

54. Jamal A, Rahman MM, Faisal M, Khan SB (2011) Studies on Photocatalytic Degradation of Acridine Orange and Chloroform Sensing Using As-Grown Antimony Oxide Microstructures. Mater Sci Appl 2: 676–683. doi: 10.4236/msa.2011.26093

55. Ye C, Meng G, Zhang L, Wang G, Wang Y (2002) A facile vapor–solid synthetic route to $Sb_2O_3$ fibrils and tubules. Chem Phys Lett 363: 34–38. doi: 10.1016/s0009-2614(02)01018-7

56. Rahman MM, Khan SB, Faisal M, Rub MA, Al-Youbi AO, et al. (2012) Determination of Olmisartan medoxomil using hydrothermally prepared nanoparticles composed $SnO_2$-$Co_3O_4$ nanocubes in tablet dosage forms. Talanta 99: 924–931. doi: 10.1016/j.talanta.2012.07.060

57. Fujii T, de-Groot FMF, Sawatzky GA, Voogt FC, Hibma T, et al. (1999) Phys Review B. 59: 3195–3202. doi: 10.1103/physrevb.59.3195
58. Rahman MM (2011) Fabrication of Mediator-free Glutamate Sensors Based on Glutamate Oxidase using Smart Micro-devices. J Biomed Nanotech 7: 351–357. doi: 10.1166/jbn.2011.1299
59. Song P, Qin HW, Zhang L, An K, Lin ZJ, et al. (2005) The structure, electrical and ethanol-sensing properties of La1−xPbxFeO3 perovskite ceramics with x≤0.3. Sens Actuator B: Chem 104: 312–316. doi: 10.1016/j.snb.2004.05.023
60. Hsueh TJ, Hsu CL, Chang SJ, Chen IC (2007) Laterally grown ZnO nanowire ethanol gas sensors. Sens Actuator B Chem 126: 473–477. doi: 10.1016/j.snb.2007.03.034
61. Tao B, Zhang J, Hui S, Wan L (2009) An amperometric ethanol sensor based on a Pd–Ni/SiNWs electrode. Sens Actuator B Chem142: 298–303. doi: 10.1016/j.snb.2009.08.004
62. Wongrat E, Pimpang P, Choopun S (2009) Comparative study of ethanol sensor based on gold nanoparticles: ZnO nanostructure and gold: ZnO nanostructure. App Surf Sci 256: 968–971. doi: 10.1016/j.apsusc.2009.02.046
63. Rahman MM, Jamal A, Khan SB, Faisal M (2011) Fabrication of chloroform sensors based on hydrothermally prepared low-dimensional $\beta$-$Fe_2O_3$ nanoparticles. Superlatt Microstruc 50: 369–376. doi: 10.1016/j.spmi.2011.07.016
64. Mujumdar S (2009) Synthesis and characterization of $SnO_2$ films obtained by a wet process. Mat Sci Poland 27: 123.
65. Rahman MM, Jamal A, Khan SB, Faisal M, Asiri AM (2011) Fabrication of Phenyl-Hydrazine Chemical Sensor based on Al-doped ZnO Nanoparticles. Sens Transduc J 134: 32–44.
66. Hagen J (1999) Heterogeneous Catalysis: Fundamentals, Wiley-VCH, Weinheim, 83–206.
67. Rahman MM, Jamal A, Khan SB, Faisal M, Asiri AM (2012) Fabrication of Highly Sensitive Acetone Sensor Based on Sonochemically Prepared As-grown $Ag_2O$ Nanostructures. Chem Engineer J 192: 122–128. doi: 10.1016/j.cej.2012.03.045
68. Sahner K, Moos R, Matam M, Tunney JJ (2005) Hydrocarbon sensing with thick and thin film p-type conducting perovskite materials, Sens. Actuator. B 108: 102–112. doi: 10.1016/j.snb.2004.12.104
69. Pokrel S, Simon CE, Quemener V, Bârsan N, Weimer U (2008) Investigation of conduction mechanism in $Cr_2O_3$ gas sensing thick films by ac impedance spectroscopy and work function changes measurements,. Sens Actuator B: Chem 133: 78–83. doi: 10.1016/j.snb.2008.01.054
70. Wang C, Fu XQ, Xue XY, Wang YG, Wang TH (2007) Surface accumulation conduction controlled sensing characteristics of p-type CuO nanorods induced by oxygen adsorption,. Nanotech 18: 145506. doi: 10.1088/0957-4484/18/14/145506
71. Choi KI, Kim HR, Kim KM, Liu D, Cao G, et al. (2010) $C_2H_5OH$ sensing

characteristics of various $Co_3O_4$ nanostructures prepared by solvothermal reaction. Sens Actuator B: Chem 146: 183–189.

72. Weng YC, Rick JF, Chou TC (2004) A sputtered thin film of nanostructured Ni/Pt/Ti on $Al_2O_3$ substrate for ethanol sensing,. Biosens Bioelectron 20: 41–51. doi: 10.1016/j.bios.2004.01.027

73. Rahman MM, Jamal A, Khan SB, Faisal M (2011) Highly sensitive ethanol chemical sensor based on Ni-doped $SnO_2$ nanostructure Materials. Biosens Bioelectron 28: 127–134. doi: 10.1016/j.bios.2011.07.024

74. Tao B, Zhang J, Hui S, Wan L (2009) An amperometric ethanol sensor based on a Pd-Ni/SiNWs electrode. Sens Actuator B Chem 142: 298–303. doi: 10.1016/j.snb.2009.08.004

75. Faisal M, Khan SB, Rahman MM, Jamal A, Asiri AM, et al. (2011) Smart chemical sensor and active photo-catalyst for environmental pollutants. Chem Engineer J 173: 178–184. doi: 10.1016/j.cej.2011.07.067

76. Weng YC, Chou TC (2002) Ethanol sensors by using $RuO_2$-modified Ni electrode. Sens Actuator B Chem 85: 246–255.

77. Khan SB, Faisal M, Rahman MM, Jamal A (2011) Exploration of $CeO_2$ nanoparticles as a chemi-sensor and photo-catalyst for environmental applications. Sci Total Environ 409: 2987–2992. doi: 10.1016/j.scitotenv.2011.04.019

78. Yang Z, Huang Y, Chen G, Guo Z, Cheng S, et al. (2009) Ethanol gas sensor based on Al-doped ZnO nanomaterial with many gas diffusing channels. Sens Actuator B Chem 140: 549–556. doi: 10.1016/j.snb.2009.04.052

79. Faisal M, Khan SB, Rahman MM, Jamal A, Umar A (2011) Ethanol chemi-sensor: Evaluation of structural, optical and sensing properties of CuO nanosheets. Mater Lett 65: 1400–1403. doi: 10.1016/j.matlet.2011.02.013

80. Rahman MM, Jamal A, Khan SB, Faisal M (2011) Fabrication of Highly Sensitive Ethanol Chemical Sensor Based on Sm-Doped $Co_3O_4$ Nanokernels by a Hydrothermal Method. J Phys Chem C 115: 9503–9510. doi: 10.1021/jp202252j

# Chapter 5

# CELL GUIDANCE ON NANOGRATINGS: A COMPUTATIONAL MODEL OF THE INTERPLAY BETWEEN PC12 GROWTH CONES AND NANOSTRUCTURES

Pier Nicola Sergi[1],IolandaMorana Roccasalvo[1].,
Ilaria Tonazzini[2], Marco Cecchini[2], Silvestro Micera[1,3]

[1]Neural Engineering Area, The BioRobotics Institute, ScuolaSuperioreSant'Anna, Pisa, Italy

[2]National Enterprise for nanoScience and nanoTechnology (NEST), IstitutoNanoscienze-National Council of Research (CNR) and Scuola Normale Superiore, Pisa, Italy

[3]Translational Neural Engineering Laboratory, Center for Neuroprosthetics& Institute of Bioengineering, School of Engineering, E´colePolytechniqueFe´de´rale de Lausanne, Lausanne, Switzerland

## ABSTRACT

### Background

Recently, the effects of nanogratings have been investigated on PC12 with respect to cell polarity, neuronal differentiation, migration, maturation of focal adhesions and alignment of neurites.

## Methodology/Principal Findings

A synergistic procedure was used to study the mechanism of alignment of PC12 neurites with respect to the main direction of nanogratings. Finite Element simulations were used to qualitatively assess the distribution of stresses at the interface between non-spread growth cones and filopodia, and to study their dependence on filopodial length and orientation. After modelling all adhesions under non-spread growth cone and filopodial protrusions, the values of local stress maxima resulted from the length of filopodia. Since the stress was assumed to be the main triggering cause leading to the increase and stabilization of filopodia, the position of the local maxima was directly related to the orientation of neurites. An analytic closed form equation was then written to quantitatively assess the average ridge width needed to achieve a given neuritic alignment ($R^2 = 0.96$), and the alignment course, when the ridge depth varied ($R^2 = 0.97$). A computational framework was implemented within an improved free Java environment (CX3D) and in silico simulations were carried out to reproduce and predict biological experiments. No significant differences were found between biological experiments and in silico simulations (alignment, $p = 0.3571$; tortuosity, $p = 0.2236$) with a standard level of confidence (95%).

## Conclusions/Significance

A mechanism involved in filopodial sensing of nanogratings is proposed and modelled through a synergistic use of FE models, theoretical equations and in silico simulations. This approach shows the importance of the neuritic terminal geometry, and the key role of the distribution of the adhesion constraints for the cell/substrate coupling process. Finally, the effects of the geometry of nanogratings were explicitly considered in cell/surface interactions thanks to the analytic framework presented in this work.

## INTRODUCTION

The outgrowth of neurites is a complex and multiscale phenomenon leading in human beings and animals to extremely specialized and effective neural structures both in the central [1] and peripheral nervous systems. Its study is fundamental to investigate the development of the nervous system, and is important in medical applications involving the regeneration of peripheral nerves after injuries [2]. Moreover, in the last decades, technological applications have made this field particularly attractive due to its implications in health care, especially for advanced prostheses for amputees [3], [4].

Although several types of neural cells have been used in the past for biological experiments, cells of pheochromocytoma cell line 12 (PC12) [5] have been widely used as a model to investigate neuronal differentiation and neuritic outgrowth. Indeed, they can reversibly adopt several neuronal characteristics upon exposure to nerve growth factor (NGF), resulting in finishing mitosis and extending protrusions, which are morphologically analogous to those of primary sympathetic neurons [6]. Biological experiments on flat substrates have shown different kinds of terminals, and "varicones" have been described when varicosities [7], [8] have been found together with growth cones (classic or non-spread), both in neuronal and PC12 cells [9],[10]. Biological experiments addressing aspects of the coupling between local extracellular topography and PC12 have also been performed using nanogratings (alternating submicron lines of ridges and grooves), which have influenced neural polarity [11], [12], cell differentiation[13], migration [14] and the modulation of focal adhesion maturation [15], [16]. Their ability to align neurites is promising for advanced biomedical applications such as the development of novel and more effective implantable neural interfaces and for improving existing solutions for peripheral nerve regeneration [17]–[22].

For this reason, over the last two decades, sophisticated tools have been developed in the field of computational neurosciences to simulate the physiology of neurons. These tools have reproduced experimental results in simple situations and have made it possible to plan new biological experiments.

Among others, CX3D (Institute of Neuroinformatics of ETH, Zurich) [23] is an open-source software written in Java used to model the growing of realistic neural networks in a three-dimensional physical space. Within this software, spheres and cylinders have mechanical properties and schematize cellular somata and neurites, allowing complex neural morphologies to be built. In addition, complex boundary conditions, as interactions among neighbouring objects and intra/extracellular diffusion, can be considered. CX3D is currently able to simulate a range of biological processes such as cellular migration and division, extension of axonal and dendritic arbours, interaction with extracellular cues, and formation of synapses. However, it currently does not allow biological contact-guidance experiments to be simulated [11], [15],[24].

The aim of this paper, therefore, was to develop a novel framework to model the outgrowth of PC12 neurites on nanogratings using an enhanced version of the CX3D source code.

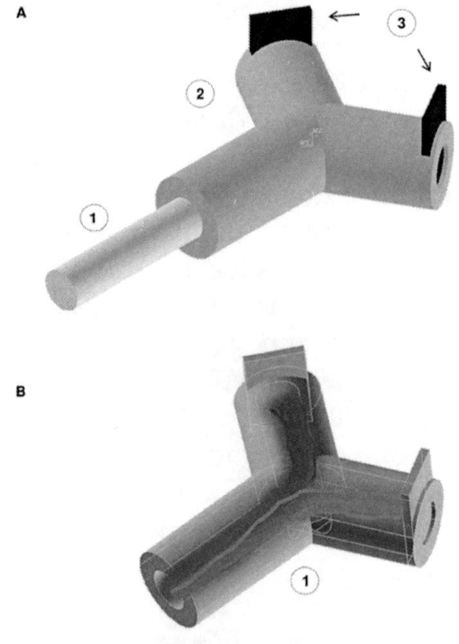


**Figure 1.** Regenerative interface (External and internal views).

(A) Scheme of reciprocal positions of (1) Healthy stump of nerve; (2) Regenerative scaffold; (3) Contacts with active sites. The healthy stump of nerve is connected with the regenerative scaffold to allow the injured axons to regenerate and contact the active sites. From these contacts, electrical signals, closely related to the patient's will of movement, can be achieved to drive neural prostheses. (B) Internal

view of regenerative scaffold with active topographic constraints (e.g. nanogratings). This kind of structure could be able to split the beam of axons improving the selectivity of contacts with the active sites. Two different populations of axons (e.g. sensory and motor) are shown in red and blue. In this concept, the beam of axon was split by the synergy of nanotopography and chemical cues.

doi:10.1371/journal.pone.0070304.g001

The whole framework was validated by comparison with biological experiments [11], and further contact-guidance in silico simulations were performed to show how this tool could optimize both the design and planning of biological experiments on nanogratings. Moreover, the results show how simulation tools combined with theoretical models could also improve the design of regenerative electrodes [17].

This particular kind of electrodes could be designed to selectively interface a high number of axons regenerating through holes, where active sites record action potentials and selectively stimulate groups of axons [17]. To enhance the selectivity of these electrodes the nervous fibres could regenerate within an active scaffold, where a synergistic action of nanotopography (e.g. nanogratings) and chemical cues can split the axon beam, allowing the separation of two families of fibres (e.g. motor and sensory fibres, see Figure 1).

## Materials and Methods

The logic flow of this study is shown in Figure 2. First, optical microscopic images were extracted from biological experiments involving PC12 growing on nanogratings. These images were analysed to characterize the real geometry of neuritic terminals. Starting from these biological data, simplified finite element (FE) models were built to reproduce the main biological features of terminals (e.g., growth cones with non-spread collapsed appearance, adhesions, etc.). In addition, FE models were used to develop an analytic equation and to implement computational simulations within CX3D (referred to further on as "in silico simulations"). Finally, the results of biological experiments were compared with those of in silico simulations to validate the whole approach and to provide predictions in more complex cases.

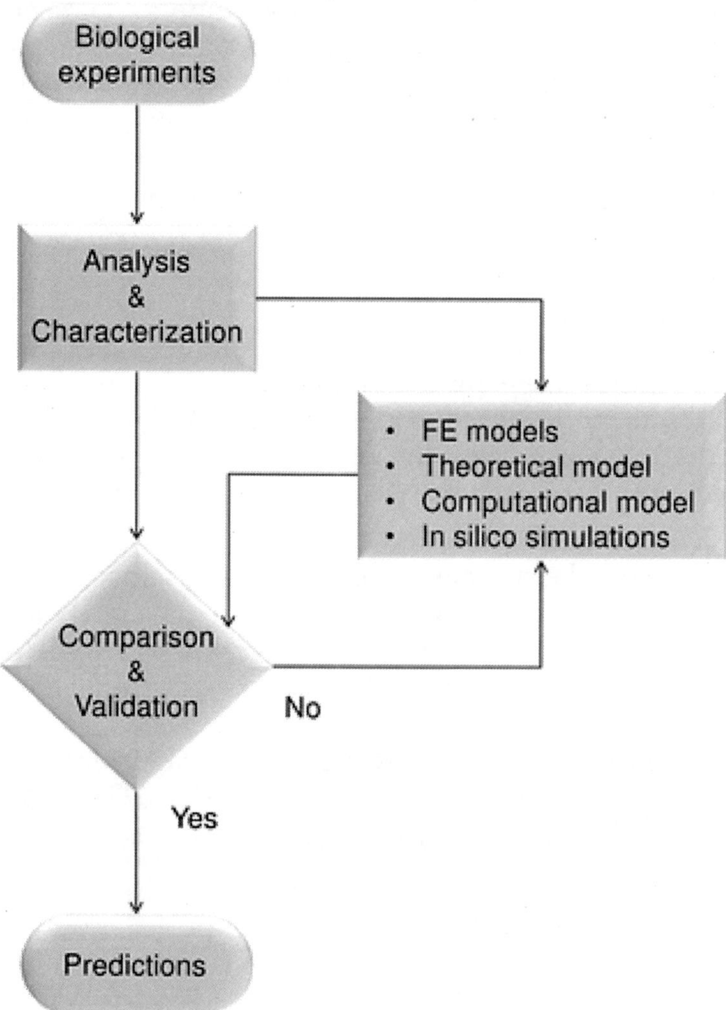


Figure 2. Logic flow of activities.

Scheme of the activities carried out in this study: images of outgrowing neurites were taken from biological experiments performed on PC12, and analysed to investigate the morphology of terminals. Simple FE models were built from these morphological data to study, accounting for geometry and constraints, the course of stress at the intersection between collapsed growth cones and filopodia. Then, an analytic model was written to account for the nanograting geometry and to implement in silico simulations. In silico results were compared with biological data to validate the whole procedure and to provide predictions on more complex geometries.

doi:10.1371/journal.pone.0070304.g002

### Cell Culture Imaging and Classification of Terminals

PC12 cells (CRL-17210, ATCC) were cultured and maintained at 37°C temperature and 5% $CO_2$ in a RMPI growth medium supplemented with 10% Horse Serum, 5% Fetal Bovine Serum, 2 mM glutamine, 10 U/ml penicillin, and 10 μM/ml streptomycin. Cells were grown on three different cyclic olefin copolymer (COC) nanogratings: ridge depth and groove width were 250 nm and 500 nm, while ridge widths were 500 nm (period 1), 1000 nm (period 1.5) and 1500 nm (period 2) respectively. Images were also obtained from cell cultures on flat surfaces. Briefly, cells were differentiated by treatment with NGF at a final concentration of 100 ng/ml on different substrates, as previously reported [15]. Differential interference contrast (DIC) images were acquired with an inverted Nikon-Ti PSF wide-field microscope (Nikon, Japan) (oil immersion 40× 1.3 NA objective - PlanFluor, Nikon) after three or four days of culture. The bright-field optical microscopy images were loaded into ImageJ (within the plug-in NeuronJ, National Institute of Health, USA). Each experiment was repeated three times independently.

In accordance with [14], only cell protrusions emerging from cell bodies and longer than 10 μm were defined as neurites and analysed. Furthermore, neuritic terminals were classified with respect to the morphology of their growth cones (spread or non-spread) [9].

Neuritic trajectories were traced and their tortuosity (defined as the ratio between the actual neurite length and its straight length from the starting point to the ending point) was measured within MatLab ® (MathWorks, Inc, USA), and reported (mean ± standard deviation) for 117 neurites growing on nanogratings with period 1 μm, at 72 h, and for n = 3 different biological experiments.

## Immunocytochemistry: Tubulin and Actin Staining

Control experiments were run to investigate and confirm the different morphologies of PC12 neuritic terminals [8] on nanogratings, by looking at their cytoskeletal composition and morphology. PC12 cells were cultured and differentiated up to 4 days with NGF (100 ng/ml) on nanogratings (ridge width 500,1000,1500 nm), as previously reported in [15].

Cells were fixed in 4% paraformaldehyde and then immunostained with anti-β Tubulin III antibody (Sigma, T2200; 6 μg/ml) and phalloidin-Alexa Fluor 647 (Invitrogen), in GDB buffer (0.2% gelatin, 0.8M NaCl, 0.5% Triton X-100, 30mM phosphate buffer, pH 7.4) [25], [26] .

Samples were then washed, incubated with Alexa Fluor 488-secondary antibody (Invitrogen) and mounted with Vectashield medium (Vector laboratories, Burlingame CA, USA).

Fluorescent samples were then examined at a TCS-SP laser scanning confocal microscope (Leica Microsystems, Germany) with a 40 x 1.4 NA objective (Plan Apochromat, Leica), and high resolution three-dimensional Z-stacks of PC12 cells were acquired. Images were then processed by ImageJ (National Institute of Health, USA).

## Finite Element Models

The majority of growth cones showed a non-spread collapsed appearance [9] on both flat and nanopatterned surfaces. Therefore, a non-spread geometry was used to implement FE models in this study.

Since tensile stress promotes the initiation of new neurites [27], stress was assumed to mainly drive the neuritic outgrowth. As a consequence, this quantity was studied to predict both direction and alignment of neurites on nanogratings, which were described through simple geometrical features such as ridge depth ($r_d$), groove width ($g_w$) and ridge width ($r_w$) (seeFigures 3A–B).

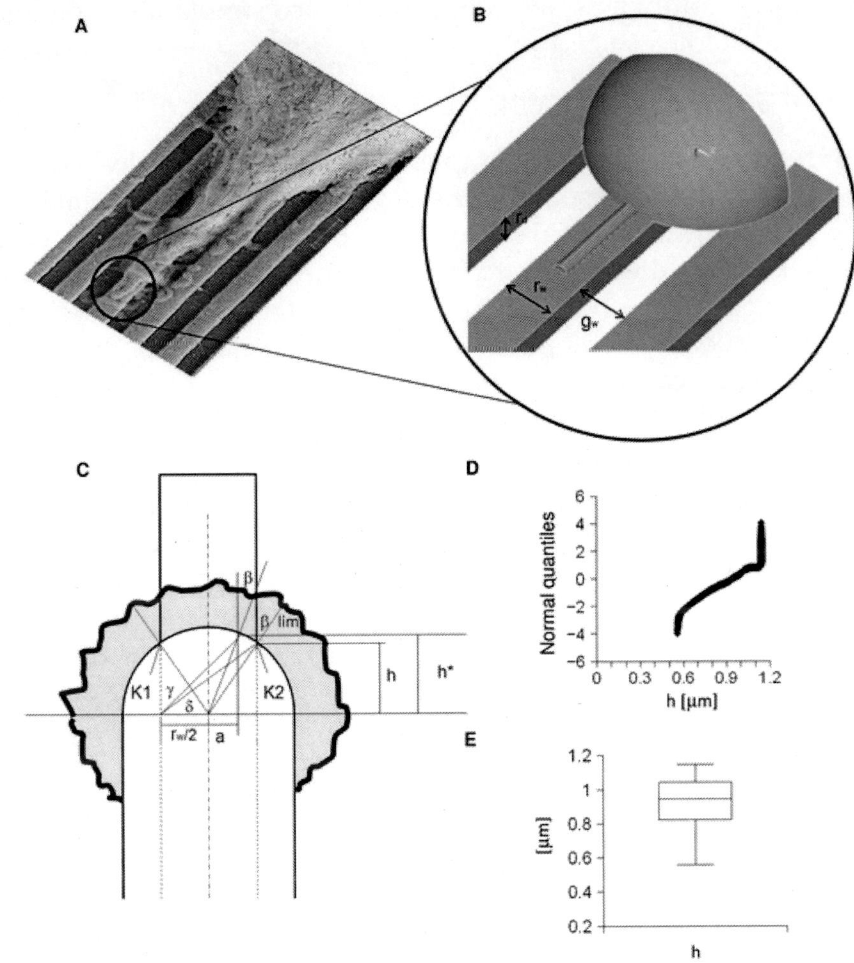


Figure 3. From biological experiments to computational models.

(A) A SEM image of filopodia emerging from a non-spread growth cone (bar = 1 µm). (B) FE model of a non-spread growth cone showing a simplified geometry together with an emerging filopodium. The set of parameters necessary to characterize the nanograting geometry is also shown: ridge width $r_w$, groove width $g_w$, and ridge depth $r_d$. (C) Bidimensional model of interactions between non-spread growth cone and ridge surface. Point K2 (together with K1, symmetric with respect to the centreline of the filopodial shaft) shows the limit angle $\beta_{lim}$. The quantity $h^*$ was connected to the actual intersection angle through a fraction of the ridge width (a). (D,E) Quantile-quantile plot of the quantity h as derived from in silico simulations, together with its box plot.

doi:10.1371/journal.pone.0070304.g003

In general, axons show a complex response to mechanical stimuli: first, a fast elastic response, then a slower passive viscoelastic behaviour [28], [29], and finally an active behaviour due to molecular motors [30]. In this work, since the protrusion-retraction cycles were faster than the outgrowth velocity of the main neuritic shaft [31], the response of filopodia was assumed to be mainly elastic. As a consequence, filopodia were characterized by two parameters (Young Modulus E = 106 Pa, and Poisson ratio ν = 0.47 [32]), and viscous effects were neglected.

In Figures 3A,B, a simple configuration, accounting for a non-spread collapsed growth cone [8],[9] with an emerging filopodium, was shown in comparison with the real geometry of a scansion electron microscopy (SEM) image. The growth cone was approximated with a quarter of sphere, while the filopodium was stylized with a half of a circular cylinder. Moreover, the growth cone was assumed to widely adhere [33], [34] to the substrate, while the filopodium was assumed to be fixed only through tip adhesions [35].

Furthermore, traction stresses, due to the reconfiguration of the neuritic cytoskeleton [36], were modelled through an imposed shortening at the contact area between the non-spread growth cone and neurite. As a consequence, the ending part of the filopodium (ending line) was totally constrained to the substrate (i.e., all degrees of freedom were set to zero), while the filopodial main shaft and the bottom surface of the growth cone were left free to shift only along their longitudinal axis.

Since any propagation of elastic waves was neglected, a quasi-static analysis of a nearly incompressible material was carried out using a three-dimensional 4-node tetrahedral structural solid (ANSYS ® Academic; Ansys, Inc. Canonsburg, Pennsylvania, USA). The mesh of volumes was achieved by using four node elements, which accounted for three translational degrees of freedom, together with the volume change rate.

To resume the local state of stress of each element, the Von Mises (VM) measure was chosen, because, within this metric, normal and shearing stresses were both considered. Indeed, VM stress was defined as:

$$\sigma_M = \left[\sigma_{xx}^2 + \sigma_{yy}^2 + \sigma_{zz}^2 - \sigma_{xx}\sigma_{yy} - \sigma_{zz}\sigma_{yy} - \sigma_{xx}\sigma_{zz} + 3(\tau_{xy}^2 + \tau_{xz}^2 + \tau_{yz}^2)\right]^{1/2},$$

where $\sigma_{xx}, \sigma_{yy}, \sigma_{zz}$ were normal stresses (respectively in x,y,z direction), while $\tau_{xy}, \tau_{xz}, \tau_{yz}$ were shearing stresses.

The previous simple configuration was used to investigate whether the maximum VM stress depended on the length and orientation of the filopodium.

In addition, a more complex configuration was considered to study filopodia emerging with different orientations: the growth cone was approximated with a quarter of sphere, while filopodia were stylized with three half circular cylinders having a length of 0.1, 0.5, 1 times the growth cone radius. This analysis was performed to compare the influence on the stress field of both filopodial length and orientation, when three different filopodia shared the same contraction of the main neuritic shaft. This case showed a strong similarity with the previous configuration accounting for a single filopodium.

In conclusion, FE simulations qualitatively provided the location of the VM stress field varying the length, number and orientation of filopodia: in both cases maximum stresses were located on the shorter and stable filopodia. As a consequence, a growth cone with non-spread collapsed appearance and one emerging filopodium was studied to obtain a bidimensional analytic model.

## From the Filopodium to the Neuritic Path

The outgrowing neurites were iteratively described as a sequence of oriented straight segments. The first tract was assumed to be an enlargement of the main filopodium, directly emerging from the collapsed growth cone and following a Laplacian distribution over the interval[11]. To account for the influence of nanotopography on the other tracts, a system of differenceequations (1) was used [37]:

$$\begin{cases} x_{t+\Delta t} = x_t + \Delta l \sin\left\{\psi\left[\beta_{t+\Delta t}(r_w, r_d, x_{t+\Delta t}, y_{t+\Delta t})\right] + \Xi\omega_t\right\} \\ y_{t+\Delta t} = y_t + \Delta l \cos\left\{\psi\left[\beta_{t+\Delta t}(r_w, r_d, x_{t+\Delta t}, y_{t+\Delta t})\right] + \Xi\omega_t\right\} \end{cases} \quad (1)$$

where $\Delta t$ was the discrete interval of time in which the neurite extended of $\Delta l$, $(x_t, y_t)$ was the geometric position of the neurite tip (corresponding to the growth cone) at time t, $\omega_t$ was a random variable with a Gaussian distribution, and $\beta_{t+\Delta t}(r_w, r_d, x_{t+\Delta t}, y_{t+\Delta t})$ was the term accounting for the angular influence of the surface geometry with respect to the actual position $(x_{t+\Delta t}, y_{t+\Delta t})$ of the growth cone at time $t+\Delta t$. Finally, $r_w$ and $r_d$ were the ridge width and depth, $\Xi = 75$, and the explicit form of the $\psi[\beta_{t+\Delta t}]$ function was expressed in Eq. (2): $\psi[\beta_{t+\Delta t}(r_w, r_d, x_{t+\Delta t}, y_{t+\Delta t})] = \mathrm{sgn}(\zeta_t)\beta_{t+\Delta t}(r_w, r_d, x_{t+\Delta t}, y_{t+\Delta t})$ (2)
where, the function $\zeta_t = (-1)^{abs[\mathrm{int}(10\omega_t)]}$ accounted for the symmetric possibility of the tip to turn on the right or on the left.

In Figure 3C, starting from the inside, two main half circles represent the border of the neuritic tip and the external shape of lamellopodia. The projections of ridges and grooves are also shown as alternated and juxtaposed rectangular areas. A symmetric ridge, having its mean line passing through the centre of the neurite, was chosen as example. Although different geometrical combinations between collapsed growth cones and ridges were possible (e.g., the contact between the non-spread growth cone and two or more ridges), more complex cases were led back to this one, because the main ridge, acting as a filter, mainly influenced the global orientation of the outgrowing neurite [31]. In the case of many principal ridges, each ridge was assumed to independently interact with the emerging filopodia, which were modelled as straight segments radiating from the centre of the growth cone.

The VM stress was assumed to promote filopodial extension, thus the most probable simulated filopodia had local VM stress maxima at the intersection with the non-spread growth cone. As a consequence, these limit filopodia passed through points K1 and K2, which were on the ridge border (see Figure 3C), and had a null length on the ridge.

Real emerging filopodia, instead, as well as having local maximum values of VM stress, laid on the surface of the main ridge. This condition further constrained their alignment to angles $\beta < \beta_{\lim}$, requiring that $h*_- > h_-$. To assess quantity $h_-$, the mean equivalent radius of 7 non-spread growth cones was measured from optical microscopy images. They were loaded within NeuronJ

plug-in to trace neuritic ends and analysed using a bidimensional CAD program. Finally, the mean equivalent radius resulted in $R_{-}$ = 1.14 μm. In particular, $h_{-}$ represented the y-coordinate of intersection between the mean collapsed growth cone (with radius $R_{-}$) and the planar projection of the ridge width. This quantity was computationally approximated (using a stylized non-spread circular growth cone with radius $R_{-}$) within CX3D (see following paragraph). In Figures 3D–E, both the distribution and the median value ( $h_{-} \approx 0.9387$) of h were shown. The simplest condition for having $\beta < \beta_{\lim}$ was chosen, so $h*_{-} = 1\,\mu m$ , and the mean value of the angle $\beta$ was expressed as: $\beta_{t+\Delta t}(r_w, r_d, x_{t+\Delta t}, y_{t+\Delta t}) = \frac{1}{2}\left[\frac{\pi}{2} - \Gamma(r_w, r_d, h*_{-})\right]$ **(3)**
where the function $\Gamma(r_w, r_d, h*_{-})$ accounted for the ridge width and depth (in micrometers). A simple form of this function was written in Eq.(4) to decouple the influence of ridge width and depth:

$$\Gamma(r_w, r_d, 1) = \frac{20}{7}\varepsilon r_d \arctan\left(\frac{1}{r_w}\right) \quad \textbf{(4)}$$

where ε was a numerical parameter, achieved through fitting of biological data [15], while $r_w = r_w(t+\Delta t)$ and $r_d = r_d(t+\Delta t)$ were geometrical parameters perceived by the growth cone in the current position.

## Implementation of CX3D Simulations

The source code of CX3D (http://www.ini.uzh.ch/projects/cx3d/) was enriched and modified to reproduce and design biological experiments on nanogratings. The standard code was improved through the addition of classes and methods to define an internal pattern (grating) accounting for biological contact-guidance experiments through Eqs. (1–4). The process of axonal guidance was assumed to be strongly affected by the growth cone dimensions, but CX3D currently lacks the classes to model the growth cone as a real physical element. To overcome this drawback, a circumference (with radius 1.14 μm, see previous paragraph) was used as a virtual growth cone to consider the interaction between growth cone and nanograting. The centre of this circumference was superimposed to the point mass of the distal segment of the neurite. As a consequence, the contacts between grating and growth cone were placed on the neurite tip and depended on their current reciprocal position. All possible interactions (on simulated gratings with period 1) were included in the following cases:

case 1: the point mass of the growth cone was placed within a groove and the growth cone overlied on two ridges. In this case, two angles ($\beta_1$ and $\beta_2$ ) quantified the influence of grating on the growth cone advancement. These angles were defined as the β angles of the segments passing near K1 and K2 (see Figure 3C). In this case the current β angle was randomly chosen between $\beta_1$ and $\beta_2$.

case 2: the point mass of the growth cone was placed within a ridge or on one of its borders. In this case, the current value of the angle β was unambiguously defined.

In both cases Eqs. (3,4) were used to calculate the current β angle, during advancement of the neurite.

A Java class was written to implement in silico simulations: the main method contained the instructions to define cell and grating geometry [11] within the CX3D physical space. In silico simulation started when two cell bodies, with a diameter of 10 μm and extending neurites, were placed on the virtual grating.

At the beginning of in silico simulation, neuritic processes extended from virtual somata to a length of 11.38 μm. According to [11] this initial length was assumed to be equal to that of neurites growing on flat surfaces. The neuritic point mass was then moved according to Eqs. (1,2), to account for the grating influence on the growing neuritic tip. The advancing speed was set equal to 20 μm/h, according to the biological data [38] for PC12 cells. When the neuritic length overcame this threshold, in silico neurites underwent extension and retraction cycles, at different speeds, until the end of CX3D simulation, according to [11].

A Java method was used to generate a length value from a Gaussian distribution with mean 18.98 μm and standard deviation of 2.65 μm. The mean of in silico distribution coincided with the mean value of biological lengths (at 12, 36, 60 hours), while the standard deviation was set equal to the smallest observed deviation. Then, for all steps, neurites cyclically extended or retracted to reach a length with a velocity derived from experimental data of cell cultures. The frequency of oscillations was implemented through a counter to reproduce the biological dynamics [11].

For each time point, biological experiments and in silico simulation had a similar range of variability. In silico neurites extended from somata into extracellular space until they reached a length that was nearly comparable with the biological values observed at the same time.

The reliability of the imposed dynamics was also studied through statistical comparison between mean experimental values of lengths (at 12, 36, 60 hours) and in silico values (at 6, 12, 18, 24, 30, 36, 42, 48, 54, 60 hours). For experimental lengths, a Shapiro-Wilk normality test resulted in $W = 0.9912$ and $p = 0.821$, while for the in silico ones in $W = 0.8951$ and $p = 0.1933$. As a consequence, both groups were assumed to have a Gaussian distribution, and a Welch t-test (two sided, independent samples) was used for statistical analysis and resulted in $p = 0.3504$.

In addition, the time courses of neuritic alignment and tortuosity were used to assess the reliability of in silico simulations with reference to biological experiments. To extract these parameters, a Java method was written to return output files containing values of virtual neurite alignment and tortuosity along time. The tortuosity was assessed using a standard procedure[37], while the alignment was calculated as the absolute value of the angle between the main direction of grating and the segment linking the starting and the ending points of the neurite.

## RESULTS

### Morphological Characterization of PC12 Terminals

Optical microscopic images of PC12 growing on three different types of gratings were analysed and compared to those of cells growing on flat substrates (Figure 4A). After three/four days of culture, two types of terminals were observed at the end of the neuritic processes: a first type had varicosities, which were more or less evident and close to the neuritic tip, where a non-spread growth cone was present; a second type ended with a spread growth cone.

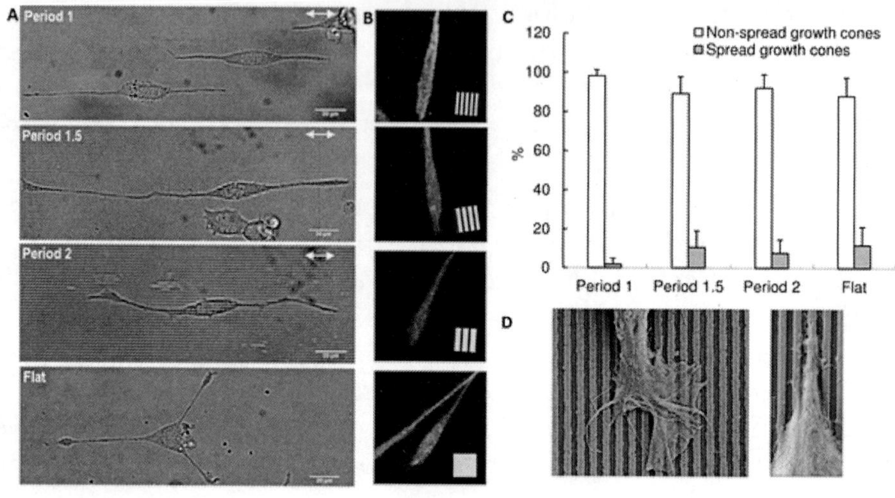


Figure 4. Analysis of neuritic terminals.

(A) Typical bright field images of PC12 cells differentiated by NGF on period 1, 1.5 and 2 gratings and on flat substrate (from the top, respectively). White arrows: grating direction; bars = 20 μm. (B) Typical confocal images of the morphological aspect of terminals with non-spread growth cones: PC12 neurite terminals grown on period 1, 1.5 and 2 gratings and on flat substrate, and stained for β3-Tubulin (green) and actin (red). Each panel side = 25 μm; square inset: grating direction. (C) Analysis of PC12 neuritic terminals over different nanogratings (periods 1, 1.5, 2 μm) and flat substrate: terminals were characterized with respect to their morphology as spread or non-spread growth cones. The analysis was carried out on 323 terminals. (D) SEM images of PC12 growth cones on period 1 nanogratings: a spread growth cone (left), magnification = 3110 X, bar length = 1 μm; a non-spread growth cone, presenting lateral transient processes (right), magnification = 12550 X, bar length = 1 μm.

doi:10.1371/journal.pone.0070304.g004

This optical analysis was supported, according to [8], by immunostaining experiments (Figure 4B), which highlighted both the morphology and cytoskeletal organization of neurite terminals on nanogratings and flat substrate. In particular, Figure 4C shows that, on flat surface, most of the observed neurites ended with non-spread growth cones (88%), while the others ended with spread growth cones (12%). Similarly, on nanogratings with period 1.5 μm, the percentage of non-spread and spread growth cones was 89% and 11% respectively, while on nanogratings with period 2 μm percentages were 92% and 8%. Finally, on period 1 μm, the percentage of non-spread growth cones increased up to 98% and spread growth cones decreased to 2%. Therefore, optical analysis supported the use of non-spread growth cones with a collapsed appearance to perform FE simulations, theoretical models and in silico simulations.

To clarify the morphological differences between the two types of terminal, two SEM images of spread (left) and non-spread (right) growth cones on period 1 nanogratings are shown in Figure 4D.

## Finite Element Models

Although the FE models were geometrically simple, they were able to qualitatively characterize the constrained contraction of the neuritic cytoskeleton. Indeed, the FE models aimed at reproducing biology, and each filopodium was fully constrained at its tip (tip adhesions) [35], while the collapsed growth cone was constrained to the substrate below (ridge surface) [33].

In order to generalize the analysis, the field of displacement was studied for non-spread collapsed growth cones with three (Figure 5A up right, left) and one (Figure 5A down right, left) emerging filopodia. This analysis was used to highlight similarities between different filopodia (Figure 5A up right, left), and similarities between FE models with three and one emerging filopodia.

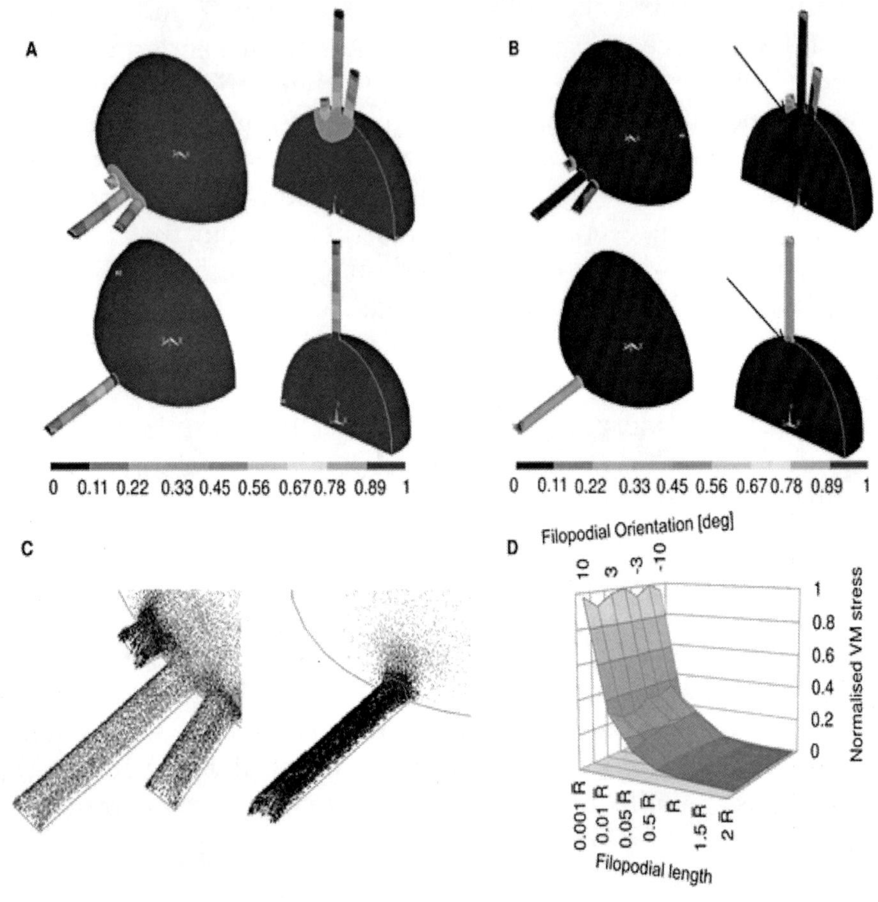


Figure 5. Finite Element models of non-spread growth cones.

(A) Displacement field for FE models of a non-spread growth cone with three (up) and one (down) emerging filopodium. The displacements were normalized over the global maximum over the whole non-spread growth cone, which accounted for the contraction of the neuritic cytoskeleton. The distribution of displacement was linear and similar among different filopodia (up). It was also similar to the displacement distribution of the model with one emerging filopodium (down). (B) Von Mises stress field for FE models of a non-spread growth cone with three (up) and one (down) emerging filopodium. VM stresses were normalized over the maximum stress at the tip of the shortest filopodium. Unlike the displacement field, the VM stresses varied among filopodia of different lengths. In particular, the course of VM along the shortest filopodium was similar for both models with three and one emerging protrusion. Arrows pointed the investigated local maxima of VM stress at the intersection between non-spread growth cones and filopodia for both models (up and down).(C) Vector plots of principal stresses for both models; left: magnification of the vector field for the model with three emerging filopodia; right: magnification of the vector field for the model with one emerging filopodium. In this case also, both fields were similar but scaled. (D) Modular surfaces accounting for the variation of intersection VM stress with filopodial orientation and length. The plot accounted for angular variation in the range (−10°, 10°) and different length of filopodia in the range $0.001R_{-}$ -$2R_{-}$ where $R_{-}$ was the radius of the non-spread growth cone. All values were normalized on the interface VM stress at 0° for a length of $0.001R_{-}$.

doi:10.1371/journal.pone.0070304.g005

In addition, the VM stress field was investigated for both models. In Figure 5B the local maxima at the intersection between filopodia and non-spread growth cone are shown using black arrows.

The similarity between the two models is also shown through the principal stress field (Figure 5C). Indeed, the shortest filopodium of the FE growth cone with three filopodia (Figure 5C left) had a vector field, which was similar, even if scaled, to that of Figure 5C right. As a consequence, the FE model with an emerging filopodium was used to investigate the course of the local maximum VM stress, varying both orientation and length of filopodia.

In Figure 5D, the values of VM stress are shown for filopodia with different orientations (−10°, 10°) and increasing lengths ranging from $0.001R_{-}$ to $2R_{-}$. All values were normalized over the intersection stress at 0° for a length of $0.001R_{-}$.

The course of data was constant with respect to the axis of orientation, showing that the VM stress was independent from this parameter. On the contrary, the course of the modular surface was highly non-linear with respect to the length axis, showing the importance of the length of the filopodium to maximize the VM stress at the intersection between non-spread growth cone and filopodia.

In conclusion, the maxima VM stresses in both models were located at the shortest and most stable (e.g., on the main ridge) filopodia.

## The Analytic Model

The local maximum VM stress, at the intersection between non-spread growth cone and filopodium, was assumed to be the main triggering cause of the directional growth of neurites on nanogratings. FE simulations showed a clear dependence of this local maximum VM stress from filopodial length. Using a simple analytic model, the nanograting geometry (e.g., ridge width and depth) was also accounted for to specify the position of the most stable filopodium.

In particular, using Eqs. (3,4), experimental data (mean alignment of neurites) were fitted [11],[15] (to quantify the parameter ε) and expressed as a function of both ridge width and depth. Furthermore, the performances of the analytic model were shown in Figure 6A to quantitatively predict the ridge width needed to achieve a given mean angle of neuritic alignment. In this plot, the previous analytic relation was inverted and the ridge width was fixed at 350 nm, to obtain the relation $r_w = r_w(\beta)$ (bold line), that was able to fit ($R^2 \cong 0.96$) experimental mean values. The $R^2$ value was the square of the correlation between the experimental data and the predicted values. This value was referred, at the same time, to all data points (mean values of experimental measurements). Then, Eq. (3) was able to model the average behaviour of 311 cells. In Figure 6A, the flat surface was considered as a limit case (vertical asymptote), when the ridge height decreased towards zero and the ridge width increased without bond. The inset showed a magnification of the analytic model performances for small values of the ridge width.

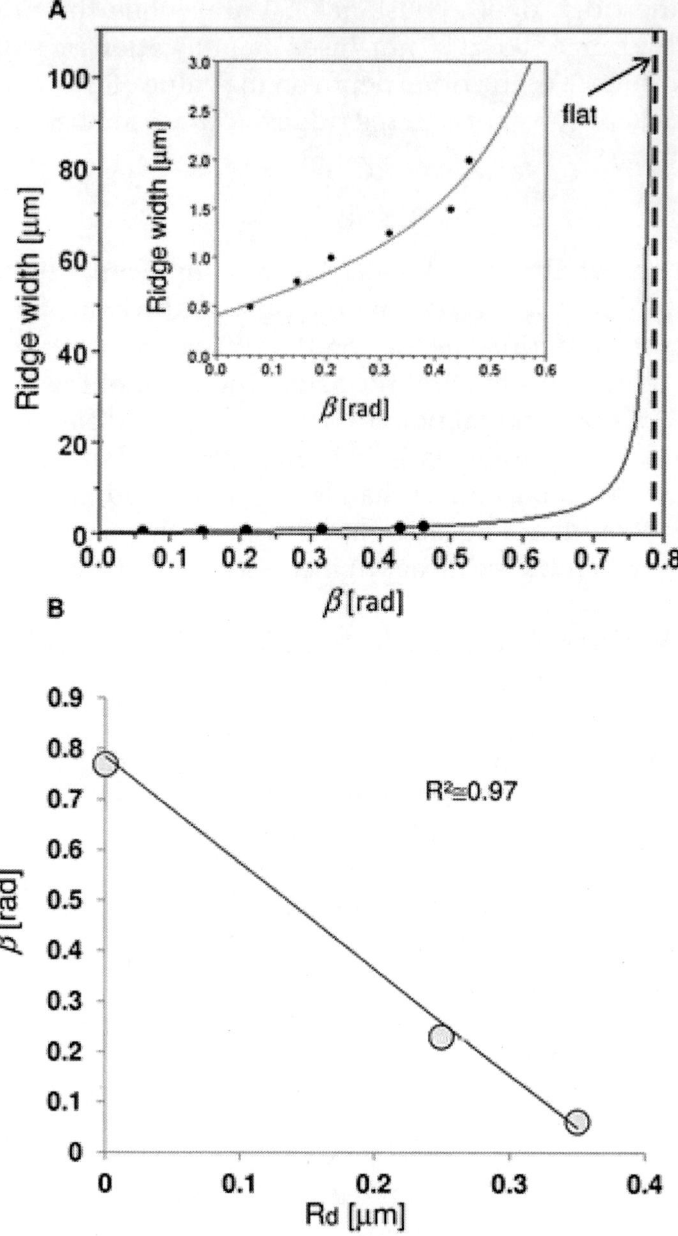


Figure 6. Analytic model.

(A) Quantitative prediction of the ridge width to achieve a given mean alignment angle β. All experimental points were obtained keeping the ridge depth constant (350 nm) while the ridge width varied in the range 500–2000 nm. Inset: magnification for small values of β. (B) Influence of the ridge depth on the value of β when the ridge width was kept constant and the ridge depth varied between 0 (flat) and 350 nm.

doi:10.1371/journal.pone.0070304.g006

Figure 6B shows that $\beta=\beta(r_w,r_d)$ was able to predict the influence of the ridge depth while keeping the ridge width constant (500 nm). The β angle was defined as the angular difference between the main direction of nanograting and the actual direction of the outgrowing neurite[11]. Experimental points corresponded to different values of the ridge depths (0 nm (flat), 250 nm, 350 nm). Furthermore, Eqs. (3,4), once the ε parameter had been quantified using equation $r_w=r_w(\beta)$, provided a negative trend near to the experimental data ($R^2 \cong 0.97$), when the ridge depth increased.

## Guided Neuritic Outgrowth: In Silico Simulations within CX3D

In Figure 7A, a DIC image, acquired with an inverted Nikon-Ti PSF wide field microscope, is shown, where PC12 cells were differentiated over a nanograting (period 1). In Figure 7B, the same biological experiment was simulated within CX3D by placing a cell population on a virtual grating and maintaining experimental parameters ($r_w = g_w = 500$ nm, $r_d = 250$ nm). In both cases, black arrows show the main orientation of the anisotropic surfaces. In Figure 6C, the comparison between biological and in silico (n = 61, mean values ± standard deviation) alignments is shown.

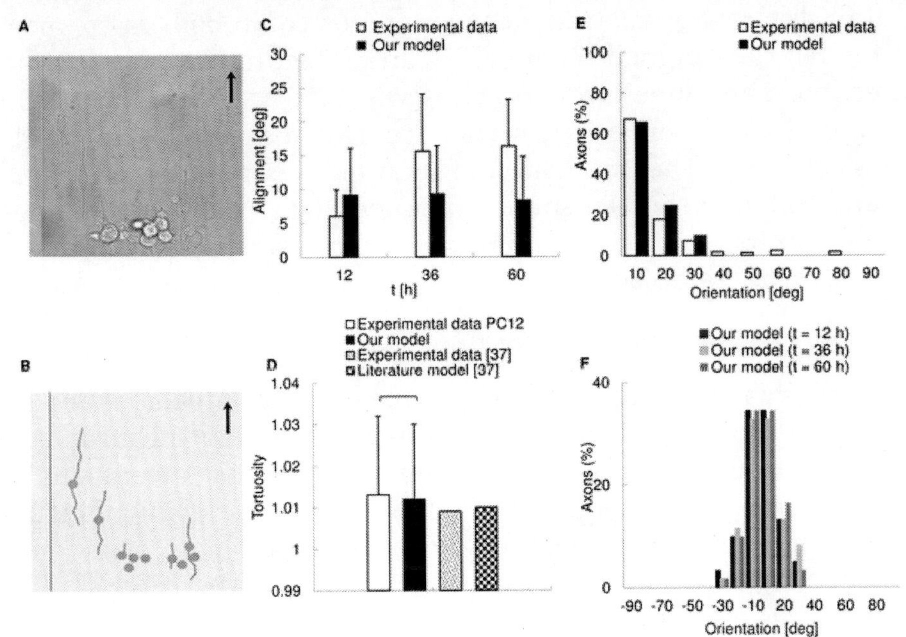


**Figure 7. In silico predictions of biological experiments.**

(A) DIC image of a cell culture acquired with an inverted Nikon-Ti PSF wide field microscope: PC12 cells were differentiated on a nanograting with period 1 and the black arrow shows the main direction of the grating. (B) Biological experiment (see A) simulated in CX3D. In silico cells were placed on a virtual nanograting with the same geometry of biological experiments and neurites grew according to the framework described by Eqs. (1–4). The black arrow indicates the main direction of the grating. (C) Comparison between biological and in silico alignments. Biological data were collected at t = 12, 36, 60 h, while in silico values were kept at t = 6, 12, 18, 24, 30, 36, 42, 48, 54, 60 h (61 neurites, error bars = standard deviations). Then, the mean biological and in silico values were compared and a Welch t-test resulted in p = 0.3571 (no significant difference). (D) Comparison between experimental, theoretical [37] and in silico mean tortuosity (117 neurites). No significant difference was found in PC12 between experimental and in silico values of tortuosity (Wilcoxon rank sum test, p = 0.2236). (E) Control of neuritic orientation for neurites on a period 1 nanograting in biological experiments [12] and in silico simulations. The percentages of neurites aligned to the grating axis within different angular ranges were reported: in both cases, most of the neurites aligned within 20° with respect to the main grating direction. The range of orientations was similar in both cases. (F) Temporal evolution of orientation control for neurites in CX3D physical space. The orientation of neurites was reported at t = 12, 36 and 60 h and showed small differences over time. Most neurites aligned to the main grating direction within small angular ranges (±20°) for any sampled time.

doi:10.1371/journal.pone.0070304.g007

At first, in silico alignment had a mean value greater than the biological one (t = 12 h), yet within the experimental error bar. Later, at t = 36 h, the mean in silico value was slightly lower than the experimental one, but again inside the range of variability. Finally, at t = 60 h, in silico alignment was lower than the biological one, and at the border of the range of variability.

Moreover, this last value was also lower than the previous ones, suggesting that in silico neurites quickly aligned with respect to grating direction over time.

A standard statistical approach was used to test whether in silico simulations were able to globally approximate biological experiments. Experimental alignments at times t = 12, 36, 60 hours, were compared with mean simulated alignments at t = 6, 12, 18, 24, 30, 36, 42, 48, 54, 60 hours. In particular, for experimental alignments, a Shapiro-Wilk normality test resulted in W = 0.8017 and p = 0.1186, while for in silico ones W = 0.9671 and p = 0.8632. Then, both groups were approximated by a normal distribution, and a Welch t-test (two sided, independent samples) resulted in p = 0.3571.

Figure 7D shows the results of experimental and in silico tortuosity obtained for PC12 cells, and [37] for primary cells. In particular, in silico neurites (n = 117) had a tortuosity of 1.012 ± 0.018, which was close to that experimentally observed (n = 117) and equal to 1.013 ± 0.019 (data were expressed as average value ± standard deviation). No significant difference was found between experimental and in silico values (Wilcoxon rank sum test, p = 0.2236).

Similarly, for primary nerve cells [37] simulations and experiments resulted in mean tortuosities of 1.01 and 1.009, respectively.

The abilty to control the neuritic orientation was shown, for a nanograting with period 1, inFigure 7E, where experimental [12] and in silico results were compared. According to biological experiments, the nanostructure led 67% of neurites to align to the main grating direction within 10° and over 90% within 30°. The other protrusions (~ 7%) were aligned within a 30°–80° range. Similarly, when the same analysis was carried out for in silico neurites growing on a period 1 virtual nanograting, most of the neurites (65%) were aligned within 10°, and over 90% within 20°, while the other cell processes (almost 10%) were oriented in a 20°–30° range with respect to the main direction of grating.

The time evolution of the orientation control is shown in Figure 7F: virtual neurites grew on a period 1 nanograting within the CX3D physical space, and the distribution of the neuritic orientations was plotted at t = 12, 36, 60 h.

First, at t = 12 h, most neurites were aligned to the grating axis within ±10° (~ 69%) and over 90% within ±20°. All the remaining protrusions were oriented within ±30°.

Then, at t = 36 h and t = 60 h, the percentages of neurites for each angle bin slightly varied. This suggested that in silico neurites had a poor variability over time in terms of angular orientation. Moreover, for any sampled time, most neurites aligned to the grating axis within small angular ranges (almost 70% within ±10° and over 90% within ±20°), according to biological experiments [12].

These results supported the use of Eqs. (3,4) to perform in silico simulations of complex interactions between nanopatterned surfaces and neuritic processes.

In Figure 8A, a contact-guidance experiment on a grating is simulated within CX3D to obtain a specific neuritic pattern (neural beam splitting). In silico cells (n = 30) were "cultivated" on a swallowtail grating, and the neurites, extending from each soma, were guided by the substrate topography along the main direction of local anisotropy. The final path of each neurite was then influenced by the actual position on the nanograting when the substrate branched into swallowtail, and neurites were able to turn on the right or on the left. Moreover, fasciculation effects were considered in CX3D standard code and cellular interactions were implemented through repulsive and attractive components: the first class prevented the overlapping of cells, the second accounted for the effect of adhesion molecules.

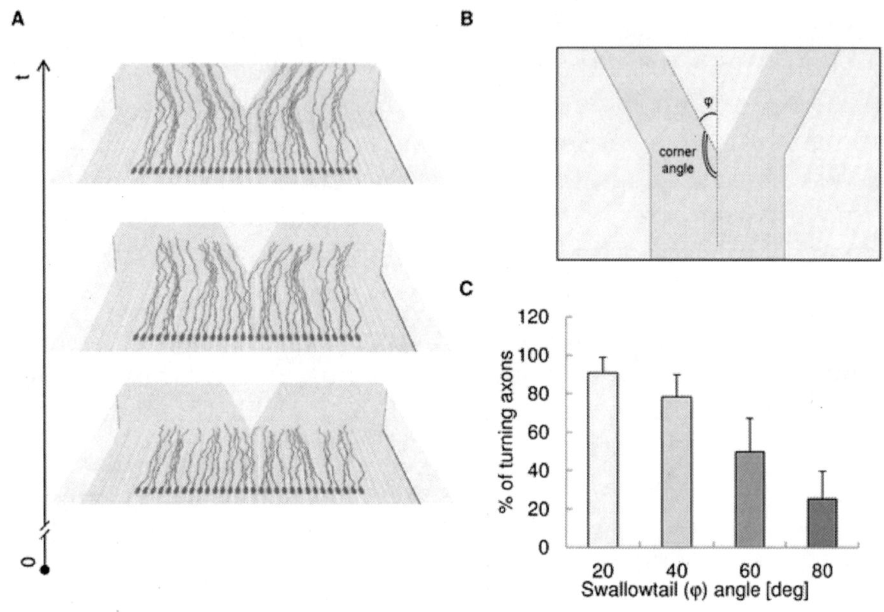


Figure 8. In silico simulations of beam splitting experiments.

(A) In silico simulation of a group of 30 cells on a swallowtail grating. Different phases of neuritic outgrowth are shown on the planes (in perspective) over time (t). In simulations, geometric and fasciculation effects were considered together. (B) Geometrical angles to model the swallowtail: the φ angle accounted for the steepness of the change between the straight grating and the following bifurcation. (C) Percentage of turning axons with different values of the φ (swallowtail) angle. This percentage decreased as the φ angle increased, in agreement with literature [39].

doi:10.1371/journal.pone.0070304.g008

As a consequence, neurites interacted with each other and modified their direction of advancement not only by following the grating influence, but also by avoiding overlaps with the near cellular protrusions and forming fascicular structures, roughly resembling axonal fascicules.

In addition, four different swallowtail geometries were implemented (with tail angles φ of 20°, 40°, 60° and 80°) to assess the influence of topography on the turning ability of the axons (Figure 8B).

On the basis of experimental results on different substrates (e.g. flat and corridors) [39]–[41], the turning ability was determined by the angle of approach between the neuritic tip and the ridge border, thus the axons that failed to turn stopped in proximity of the corner.

In silico results were shown in Figure 8C: the axons were able to turn on all tested geometries, but the percentages of turning axons were ~ 91%, ~ 78%, ~ 50% and ~ 25% for tail angles of 20°, 40°, 60° and 80° respectively. These results showed that the turning frequency decreased as the tail angle increased (and the corner angle decreased), accordingly to the experimental results reported in literature [39].

## DISCUSSION

PC12 neurites growing on nanogratings and flat substrates were investigated through optical analysis of DIC and SEM images, further strengthened by immunocytochemistry. As a result, most of the growth cones showed a non-spread collapsed appearance (Figure 4B).

Nevertheless, non-spread growth cones were regularly involved in pathfinding through a refined sensing of substrates, consistently with biological experiments on nanopatterned [13], [15],[16] and flat substrates [8], [9].

In particular, since nanogratings induce a high degree of anisotropy, the scanning mechanism of the filopodia highly depended on their current position with respect to the principal direction of nanopatterns, and the filopodia were grouped together in two different subsets: non-aligned and aligned [31]. Indeed, only filopodia entirely extending over the ridge surfaces had a continuum contact zone and enlarged through a robust F-actin network [31].

Although this simple mechanism seemed to explain the reason why the neurites tended to avoid unaligned tracks, unfortunately it was not able to quantify the mean neuritic misalignment[11], nor its changes related to the dimensions of nanograting [15]. As a consequence, this work aimed at quantifying both these issues starting from the reciprocal position between non-spread growth cones and ridge surface.

The local maximum VM stress, at the interface between filopodia and non-spread growth cone, was assumed to be the main cause triggering the biophysical reactions leading to the outgrowth of neurites. This mechanism had the easiness of a "search and capture" mechanism[31], but a much greater ability to orient the path of the neurites. Accordingly, the size of non-spread growth cones, the dimensions of ridges, together with the characteristics of growth cone/substrate coupling, emerged as the main factors involved in neuritic pathfinding.

Furthermore, advanced issues, such as the balance between deterministic and stochastic effects [42], also emerged as being fundamental to the building of the neuritic paths.

This framework was able to implement direct and inverse in silico simulations with a chosen number of neurites and with single or combined substrates of any complexity. Direct simulations, as previously shown in the results section, were calibrated through experimental data to approximately reproduce the biological behaviour of PC12 neurites.

However, it was difficult to clearly define what "to reproduce biological experiments" generally means: "optical" likeness on its own (as shown in Figures 7A–B) was necessary but not sufficient to provide suitable in silico simulations.

As a consequence, three further parameters were chosen in this work to strengthen the concept of likeness. The first input parameter was the dynamics of the neuritic outgrowth. In silico dynamics globally approximated the experimental one with a standard level of confidence (Welch t-test, $p = 0.3504$).

The second output parameter was the mean angle of alignment. Also in this case, statistical analysis showed the absence of significant differences (Welch t-test, $p = 0.3571$) between biological experiments and in silico simulations over the global time range ($t = 6 \div 60$ h from NGF administration), and for a standard level of confidence (95%).

Moreover, Figure 7E shows that most of the simulated neurites aligned to the main grating direction within 20° according to the biological data (respectively the 90% and 85%). Furthermore, in both biological experiments and in silico simulations, the value of 30° seemed to represent a cut-off value, over which the percentage of growing neurites was negligible (biological experiments) or null (in silico simulations).

The third output parameter was tortuosity, and quantitative comparisons were made with PC12 and primary cells [37]. In silico results were in good agreement not only with experimental values for both PC12 (difference ~ 0.1%) and primary cells (difference ~ 0.3%) [37], but also with the theoretical simulation of primary cells (difference ~ 0.2%).

Therefore, the similarity between in silico and biological neurites was assessed using all these parameters: once the dynamics of neurites and the geometry of nanograting were inserted within in silico simulations, the angles of alignment and tortuosity were also in agreement with biological data.

This seemed to suggest that the global outputs (alignment, tortuosity) resulting from the coupling between dynamics and topography (inputs) were able to catch the main features of the real PC12 outgrowth on nanograting. As a consequence, this simple system, ruled by deterministic interactions and stochastic fluctuations, was able to approximate a much more complex cellular system.

In this study, both dynamics and angles of alignment were analysed along the whole interval of time, so the likeness was globally computed for all in silico simulations. This procedure was effective because the length of PC12 neurites had only small changes along the NGF phase (about 10 μm). Nevertheless, in the case of greater elongations in time and large variations in alignment, global likeness should be replaced by local likeness, and in silico values very close to experimental ones may be required for each experimental time point both for input dynamics and output alignments.

In conclusion, this work presented a synergistic procedure, which is a promising starting point towards more complex simulations accounting for also highly irregular dynamics and inhomogeneous outgrowth. Thanks to its simplicity, this procedure can be used to implement inverse simulations coupled with different computational strategies (e.g., genetic algorithms [43]) and to approach a set of complex optimization problems, often with many possible solutions.

Indeed, just starting from a given set of optimal neuritic paths, many suitable surfaces respecting initial constraints and boundary conditions can be identified. Then, from this set, the best surface has to be found to optimize technological issues and time variant effects.

In addition, through inverse simulations, the specific behaviour of cells (PC12 or primary cells with a similar behaviour at the cell/ surface length scale) can be inserted as "biological constraint" into design procedures together with other classic issues (e.g., surface technology, biocompatibility of materials).

This computational model could be also applied to dynamic problems related to the effectiveness of regenerative interfaces [17]: in this case, mechanical interferences between afferent (sensory), efferent (motor), and autonomic fibres largely decrease the reliability of the coupling between axons and neural interface. At first, indeed, many contacts are established between motor fibres and active borders of holes lying over the partition surface. Then, sensory fibres grow faster than motor fibres, causing compressive interferences within the holes. Since compressive stresses choke motor fibres, most of the remaining connections are with sensory fibres: this precludes access to the motor functions of patients.

A possible way to decrease these effects is to split motor and sensory fibres, conveying them to different partition surfaces (see Figure 1). To this aim, the dynamic characteristics of fibres (e.g., velocity of growth, degree of superposition, ability to branch, interaction forces between neurites) could be synergistically used together with static features (e.g., geometry and elasticity) within the presented framework. Finally, all these possibilities can be coupled with the action of chemical gradients [44], deriving from any set of sources in a chemical active environment, and making it possible to study the non linear superposition between chemical and topographic cues both in static and dynamic conditions.

## Author Contributions

Wrote the paper: PNS IMR IT MC SM. Conceived and performed experiments: IT MC. Performed optical analyses: IMR IT. Conceived

FE simulations and wrote analytic and computational models: PNS. Improved the CX3D code and performed in silico simulations: IMR. Performed statistical analyses: PNS. Approved the final version of the manuscript: PNS IMR IT MC SM.

## REFERENCES


1. Arimura N, Kaibuchi K (2007) Neuronal polarity: from extracellular signals to intracellular mechanisms. Nat Rev Neurosci 8: 194–205. doi: 10.1038/nrn2056
2. Girard C, Liu S, Cadepond F, Adams D, Lacroix C, et al. (2008) Etifoxine improves peripheral nerve regeneration and functional recovery. Proc Natl Acad Sci U S A 105: 20505–20510. doi: 10.1073/pnas.0811201106
3. Carrozza MC, Cappiello G, Micera S, Edin BB, Beccai L, et al. (2006) Design of a cybernetic hand for perception and action. Biol Cybern 95: 629–644. doi: 10.1007/s00422-006-0124-2
4. Cipriani C, Controzzi M, Carrozza MC (2011) The SmartHand transradial prosthesis. J Neuroeng Rehabil 8: 29. doi: 10.1186/1743-0003-8-29
5. Greene LA, Tischler AS (1976) Establishment of a noradrenergic clonal line of rat adrenal pheochromocytoma cells which respond to nerve growth factor. Proc Natl Acad Sci U S A 73: 2424–2428. doi: 10.1073/pnas.73.7.2424
6. Foley JD, Grunwald EW, Nealey PF, Murphy CJ (2005) Cooperative modulation of neuritogenesis by PC12 cells by topography and nerve growth factor. Biomaterials 26: 3639–3644. doi: 10.1016/j.biomaterials.2004.09.048
7. Aletta JM, Greene LA (1988) Growth cone configuration and advance: a time-lapse study using video- enhanced differential interference contrast microscopy. J Neurosci 8: 1425–1435.
8. Mingorance-Le Meur A, Mohebiany AN, O'Connor TP (2009) Varicones and growth cones: two neurite terminals in PC12 cells. PLoS One 4: e4334. doi: 10.1371/journal.pone.0004334
9. Dail M, Richter M, Godement P, Pasquale EB (2006) Eph receptors inactivate R-Ras through different mechanisms to achieve cell repulsion. J Cell Sci 119: 1244–1254. doi: 10.1242/jcs.02842
10. Connolly JL, Seeley PJ, Greene LA (1987) Rapid regulation of neuronal growth cone shape and surface morphology by nerve growth factor. Neurochem Res 12: 861–868. doi: 10.1007/BF00966307
11. Ferrari A, Cecchini M, Serresi M, Faraci P, Pisignano D, et al. (2010) Neuronal polarity selection by topography-induced focal adhesion control. Biomaterials 31: 4682–4694. doi: 10.1016/j.biomaterials.2010.02.032
12. Cecchini M, Bumma G, Serresi M, Beltram F (2007) PC12 differentiation on biopolymer nanostructures. Nanotechnology 18: 1–7. doi: 10.1088/0957-4484/18/50/505103

13. Ferrari A, Faraci P, Cecchini M, Beltram F (2010) The effect of alternative neuronal differentiation pathways on PC12 cell adhesion and neurite alignment to nanogratings. Biomaterials 31: 2565–2573. doi: 10.1016/j.biomaterials.2009.12.010

14. Ferrari A, Cecchini M, Degl'Innocenti R, Beltram F (2009) Directional PC12 cell migration along plastic nanotracks. IEEE Trans Biomed Eng 56: 2692–2696. doi: 10.1109/TBME.2009.2027424

15. Ferrari A, Cecchini M, Dhawan A, Micera S, Tonazzini I, et al. (2011) Nanotopographic control of neuronal polarity. Nano Lett 11: 505–511. doi: 10.1021/nl103349s

16. Meucci S, Tonazzini I, Beltram F, Cecchini M (2012) Biocompatible noisy nanotopographies with specific directionality for controlled anisotropic cell cultures. Soft Matter 8: 1109–1119. doi: 10.1088/0957-4484/18/50/505103

17. Navarro X, Krueger TB, Lago N, Micera S, Stieglitz T, et al. (2005) A critical review of interfaces with the peripheral nervous system for the control of neuroprostheses and hybrid bionic systems. J Peripher Nerv Syst 10: 229–258. doi: 10.1111/j.1085-9489.2005.10303.x

18. Rossini PM, Micera S, Benvenuto A, Carpaneto J, Cavallo G, et al. (2010) Double nerve intraneural interface implant on a human amputee for robotic hand control. Clin Neurophysiol 121: 777–783. doi: 10.1016/j.clinph.2010.01.001

19. Dario P, Garzella P, Toro M, Micera S, Alavi M, et al. (1998) Neural interfaces for regenerated nerve stimulation and recording. IEEE Trans Rehabil Eng 6: 353–363. doi: 10.1109/86.736149

20. Stieglitz T, Ruf HH, Gross M, Schuettler M, Meyer JU (2002) A biohybrid system to interface peripheral nerves after traumatic lesions: design of a high channel sieve electrode. Biosens Bioelectron 17: 685–696. doi: 10.1016/S0956-5663(02)00019-2

21. Micera S, Rossini PM, Rigosa J, Citi L, Carpaneto J, et al. (2011) Decoding of grasping information from neural signals recorded using peripheral intrafascicular interfaces. J Neuroeng Rehabil 8: 53. doi: 10.1186/1743-0003-8-53

22. Micera S, Citi L, Rigosa J, Carpaneto J, Raspopovic S, et al. (2010) Decoding information from neural signals recorded using intraneural electrodes: toward the development of a neurocontrolled hand prosthesis. Proc IEEE 98: 407–417. doi: 10.1109/jproc.2009.2038726

23. Zubler F, Douglas R (2009) A framework for modeling the growth and development of neurons and networks. Front Comput Neurosci 3: 25. doi: 10.3389/neuro.10.025.2009

24. Clark P, Connolly P, Curtis AS, Dow JA, Wilkinson CD (1990) Topographical control of cell behaviour: II. Multiple grooved substrata. Development 108: 635–644. doi: 10.1109/jproc.2009.2038726

25. Wieringa P, Tonazzini I, Micera S, Cecchini M (2012) Nanotopography induced contact guidance of the F11 cell line during neuronal differentiation:

a neuronal model cell line for tissue scaffold development. Nanotechnology 23: 275102. doi: 10.1088/0957-4484/23/27/275102

26. Borgesius NZ, van Woerden GM, Buitendijk GHS, Keijzer N, Jaarsma D, et al. (2011) βCaMKII plays a nonenzymatic role in hippocampal synaptic plasticity and learning by targeting αCaMKII to synapses. J Neurosci 31: 10141–10148. doi: 10.1523/JNEUROSCI.5105-10.2011

27. Zheng J, Lamoureux P, Santiago V, Dennerll T, Buxbaum RE, et al. (1991) Tensile regulation of axonal elongation and initiation. J Neurosci 11: 1117–1125.

28. Dennerll TJ, Joshi HC, Steel VL, Buxbaum RE, Heidemann SR (1988) Tension and compression in the cytoskeleton of PC-12 neurites. II: quantitative measurements. J Cell Biol 107: 665–674. doi: 10.1083/jcb.107.2.665

29. Lamoureux P, Buxbaum RE, Heidemann SR (1989) Direct evidence that growth cones pull. Nature 340: 159–162. doi: 10.1038/340159a0

30. Bernal R, Pullarkat PA, Melo F (2007) Mechanical properties of axons. Phys Rev Lett 99: 018301. doi: 10.1103/PhysRevLett.99.018301

31. Jang KJ, Kim MS, Feltrin D, Jeon NL, Suh KY, et al. (2010) Two distinct filopodia populations at the growth cone allow to sense nanotopographical extracellular matrix cues to guide neurite outgrowth. PLoS One 5: e15966. doi: 10.1371/journal.pone.0015966

32. Betz T, Koch D, Lu YB, Franze K, Käs JA (2011) Growth cones as soft and weak force generators. Proc Natl Acad Sci U S A 108: 13420–13425. doi: 10.1073/pnas.1106145108

33. Arregui CO, Carbonetto S, McKerracher L (1994) Characterization of neural cell adhesion sites: point contacts are the sites of interaction between integrins and the cytoskeleton in PC12 cells. J Neurosci 14: 6967–6977.

34. Renaudin A, Lehmann M, Girault J, McKerracher L (1999) Organization of point contacts in neuronal growth cones. J Neurosci Res 55: 458–471. doi: 10.1002/(SICI)1097-4547(19990215)55:4<458::AID-JNR6>3.0.CO;2-D

35. Steketee MB, Tosney KW (2002) Three functionally distinct adhesions in filopodia: shaft adhesions control lamellar extension. J Neurosci 22: 8071–8083.

36. Baas PW, Ahmad FJ (2001) Force generation by cytoskeletal motor proteins as a regulator of axonal elongation and retraction. Trends Cell Biol 11: 244–249. doi: 10.1016/S0962-8924(01)02005-0

37. Borisyuk R, Cooke T, Roberts A (2008) Stochasticity and functionality of neural systems: mathematical modelling of axon growth in the spinal cord of tadpole. Biosystems 93: 101–114. doi: 10.1016/j.biosystems.2008.03.012

38. Goodhill GJ, Gu M, Urbach JS (2004) Predicting axonal response to molecular gradients with a computational model of filopodial dynamics. Neural Comput 16: 2221–2243. doi: 10.1162/0899766041941934

39. Francisco H, Yellen BB, Halverson DS, Friedman G, Gallo G (2007) Regulation of axon guidance and extension by three-dimensional constraints. Biomaterials 28: 3398–3407. doi: 10.1016/j.biomaterials.2007.04.015

40. Bouquet C, Soares S, von Boxberg Y, Ravaille-Veron M, Propst F, et al. (2004) Microtubule-associated protein 1B controls directionality of growth cone migration and axonal branching in regeneration of adult dorsal root ganglia neurons. J Neurosci 24: 7204–7213. doi: 10.1523/JNEUROSCI.2254-04.2004
41. Burmeister DW, Goldberg DJ (1988) Micropruning: the mechanism of turning of Aplysia growth cones at substrate borders in vitro. J Neurosci 8: 3151–3159.
42. Maskery S, Shinbrot T (2005) Deterministic and stochastic elements of axonal guidance. Annu Rev Biomed Eng 7: 187–221. doi: 10.1146/annurev.bioeng.7.060804.100446
43. Ciofani G, Sergi PN, Carpaneto J, Micera S (2011) A hybrid approach for the control of axonal outgrowth: preliminary simulation results. Med Biol Eng Comput 49: 163–170. doi: 10.1007/s11517-010-0687-x
44. Ciofani G, Raffa V, Menciassi A, Cuschieri A, Micera S (2009) Magnetic alginate microspheres: system for the position controlled delivery of nerve growth factor. Biomed Microdevices 11: 517–527. doi: 10.1007/s10544-008-9258-4

# Chapter 6

# A SUPER ENERGY MITIGATION NANOSTRUCTURE AT HIGH IMPACT SPEED BASED ON BUCKYBALL SYSTEM

Jun Xu , Yibing Li , Yong Xiang [1, 2 ,3*], Xi Chen[1,4,5]

[1]Columbia Nanomechanics Research Center, Department of Earth and Environmental Engineering, Columbia University, New York, New York, United States of America

[2]State Key Laboratory of Automotive Safety & Energy, Department of Automotive Engineering, Tsinghua University, Beijing, China

[3]State Key Lab of Electronic Thin Films and Integrated Devices, School of Energy Science and Engineering, University of Electronic Science and Technology of China, Chengdu, Sichuan, China

[4]Department of Civil and Environmental Engineering, Hanyang University, Seoul, Korea, 5International Center for Applied Mechanics, SV Lab, Xi'an Jiaotong University, Xi'an, China

## ABSTRACT

The energy mitigation properties of buckyballs are investigated using molecular dynamics (MD) simulations. A one dimensional buckyball long chain is employed as a unit cell of granular fullerene particles. Two types of buckyballs i.e. $C_{60}$ and $C_{720}$ with recoverable and non-recoverable behaviors are chosen respectively. For $C_{60}$ whose deformation is relatively small, a dissipative contact model is proposed. Over 90% of the total impact energy is proven to be mitigated through interfacial reflection of wave propagation, the van der Waals interaction, covalent potential energy and atomistic kinetic energy evidenced by the decent force attenuation and elongation of transmitted impact. Further, the $C_{720}$ system is found to outperform its $C_{60}$ counterpart and is able to mitigate over 99% of the total kinetic energy by using a much shorter chain thanks to its non-recoverable deformation which enhances the four energy dissipation terms. Systematic studies are carried out to elucidate the effects of impactor speed and mass, as well as buckyball size and number on the system energy mitigation performance. This one dimensional buckyball system is especially helpful to deal with the impactor of high impact speed but small mass. The results may shed some lights on the research of high-efficiency energy mitigation material selections and structure designs.

## INTRODUCTION

Protection of materials and devices under high-speed impact, whose most critical task is energy mitigation and absorption [1]–[3], poses a major challenge in engineering. For ballistic loading, i.e. high impact speed with small impact mass, the force attenuation should be the priority [4] to effectively mitigate impact energy. Woven fabric composites [5]–[8], sandwich structure [9]–[11], metal foams [12]–[14] and nanomaterials [15]–[21] are widely used for energy mitigation upon high speed impact, which primarily consume the impact energy through widespread failure or extensive deformation.

Granular material arranging in a chain-like structure [22], [23] is attractive for force attenuation, and such a discrete system effectively responds to impact loading via stress wave propagation across various interfaces to reduce the transmitted force. Pioneering work on the characteristics of the solitary wave propagation in a homogeneous chain of metallic spheres based on the Hertz contact law was established by Nesterenko [24]. Since then, many contributions have been put forward to refine the chain system for outstanding energy damping ability, including the material and geometrical parameters [25], [26], arrangements [27], [28], and model parameterizations of different granular materials [29]–[31].

Recently, with the development of nanomaterial, carbon nanotubes (CNTs) [21], [32] have been one of the promising candidates for impact energy absorption thanks to its ultra-high modulus and strength [33]–[35]. Buckyballs, another branch of fullerene family, also have high potential for energy mitigation owing to their excellent mechanical properties and unique morphology[36], [37]. According to our previous work [20], [38], the progressive buckling and densification in response to impact loading, as well as the particular non-recoverable portraits of larger buckyballs, may help to dissipate and absorb intense stress waves. Thus, inspired by granular materials, it is envisioned that the stacking of nano-sized buckyballs could exhibit excellent energy mitigation capabilities.

In this paper, two representative buckyballs $C_{60}$ and $C_{720}$ stacked in one-dimensional chain-like system are chosen to study the mechanical behavior subject to high speed impact. For the small $C_{60}$ buckyball chain, an analytical model based on the Hertz contact law is suggested by analogy to the fundamental Nesterenko's model. Molecular dynamics (MD) simulations are employed to study the transmitted force history and the peak force attenuation. Stress wave propagation characteristics are also investigated such that system effective response is evaluated. For the giant $C_{720}$ buckyball chain, MD simulations are used to compute the contact forces on the impactor and receiver, as well as the stress wave propagation. Further, the effect of the impact mass and speed on the system performance is thoroughly studied to fully unveil the energy mitigation mechanism. Finally, buckyballs with various sizes are embedded into the chain system to explore the particle size effect on the energy dissipation ability.

## COMPUTATIONAL MODEL AND METHOD

Small and large buckyballs behave differently upon impact: the smaller ones are often resilient while the larger ones exhibit non-recovery phenomenon after unloading [38]. In this study, $C_{60}$and $C_{720}$ are selected to represent "recoverable buckyball" and "non-recoverable buckyball" respectively. In continuum modeling, buckyballs are assumed to share the same effective Young's modulus $E = 5$ TPa and nominal wall thickness $t = 0.66$ nm [38]. The densities of $C_{720}$ and $C_{60}$ are $\rho_{C_{720}} = 1.975\ \text{kg/m}^3$ and $\rho_{C_{60}} = 5.455\ \text{kg/m}^3$ respectively. The other basic physical parameters of $C_{720}$ and $C_{60}$ are listed in Ref. [20]. To simulate a granular system, we assume the identical buckyballs are packed in a simple cubic manner such that the stress wave would be confined within one dimension (effects caused by different packaging arrangements have been discussed in Ref [38]). We have shown that the system deformation mode and the energy absorption/mitigation ability are independent of the arrangement number in both vertical and horizontal lineups in previous work [38]. In addition, preliminary simulation also reveals that system with multi-column stacking has no obvious difference in deformation behavior and unit energy absorption rate.

Thus, by taking advantage of symmetry, a long chain of buckyball system is simulated. The "long chain" is set to be at least 20 times in length than its width, and a typical system contains 100 buckyballs. The computational cell is illustrated in Figure 1, where the buckyball system subjects to the impact of a rigid left plate with incident energy $E_{\text{impactor}}$ and the impact speed is varied from 100 m/s to 1000 m/s which is conventionally considered as high impact speed domain, mainly aiming at the ballistic impact related problem.

Mass changing falls into the domain where the maximum strain is large enough while the temperature rising of the buckyball caused by the kinetic energy is below 800 K when buckyball may remain stable. A rigid and fixed right plate serves as a receiver which would indicate the energy mitigation capability of the protective system (the buckyball chain is sandwiched between the plates). Force histories on the left and right plates are recorded.

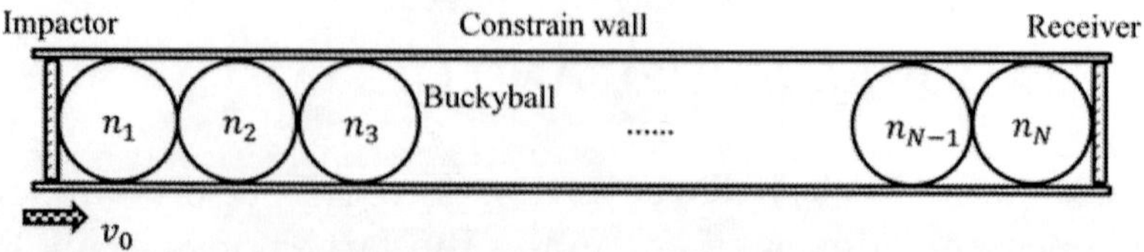


Figure 1. Illustration of one-dimensional buckyball chain setup as an impact protector.

doi:10.1371/journal.pone.0064697.g001

A full atomistic description of the buckyball is used. MD simulation is performed based on LAMMPS (large-scale atomic/molecular massively parallel simulator) platform with the NVE ensemble (micro-canonical ensembles) [39] after running initial equilibrium. A pairwise Lennard-Jones (L-J) potential term is added to the buckyball potential to account for the steric and van der Waals carbon–carbon interactions $U(r_{ij}) = 4\varepsilon_{CC}\left[(\sigma_{CC}/r_{ij})^{12} - (\sigma_{CC}/r_{ij})^{6}\right]$ where $\varepsilon_{CC}$ is the depth of the potential well between carbon-carbon atoms; $\sigma_{CC}$ is the finite distance where the carbon-carbon potential is zero; $r_{ij}$ is the distance between the two carbon atoms. Here, L-J parameters for the carbon atoms of the buckyball are $\sigma_{CC} = 3.47\text{Å}$ and $\varepsilon_{CC} = 0.27647\ \text{kJ/mol}$ as used in the original parameterization of Girifalco [40] and Van der Waals interaction governs in the plate-buckyball interaction. Carbon atoms are employed to make both the impactor and receiver plates with varying masses in the following simulation to set various loading conditions (varying impactor mass) while the interactions between the plates and buckyballs remain as carbon-carbon ones. Details of the simulation methods are described elsewhere [38]. To simulate the long one dimensional chain, four L-J walls with the same parameters are set as four sides of the simulation box to provide necessary lateral constraints from simple cubic packing. A time integration step of 1 fs is used and periodical boundary conditions are applied in the *x,y* plane to eliminated the boundary effect.

## REPRESENTATIVE IMPACT BEHAVIOR

### Dynamic response of $C_{60}$ chain system

#### *Hertzian model*

Interactions between particles in the one-dimensional chain system subject to contact loading may be treated based on the Hertz law [24]. Similar to granular particles, each $C_{60}$ molecule in the chain system undergoes relatively small deformation without any buckling or bifurcation. In addition, the characteristic time

$$\tau \approx 10^{-1} \sim 10^{0}\,\text{ns} \gg T \approx 2.5R_{C_{60}}/c_1 \approx 5.71 \times 10^{-5}\,\text{ns}$$

where $R_{C_{60}}$ is the radius of $C_{60}$ and $c_1 = \sqrt{E/\rho_{C_{60}}}$ is the wave speed [24]. Therefore, the Hertz contact law still approximately holds for the dynamic response of $C_{60}$ chain system.

Consider a one dimensional chain of $N$ same $C_{60}$s with mass $m_{C_{60}}$ , radius $R_{C_{60}}$ and Young's modulus $E$ and Poisson's ratio $\nu$ in contact without any precompression. The Hertzian contact law between neighboring buckyballs and could be expressed as[41]

$$F = k_c\delta^{\frac{3}{2}} = \frac{4}{3}E^{*}\sqrt{R^{*}}\cdot\left(2R_{C_{60}} - (x_2 - x_1)\right)^{\frac{3}{2}} \quad \textbf{(1)}$$

where $F$ is the contact force, $k_c$ referring the elastic coefficient, $\delta$ is deformation, $x_2$, $x_1$ are coordinates of two neighboring buckyball centers ($x_2 > x_1$); $E^{*} = E/2(1-\nu^2)$, and $R^{*} = R_{C_{60}}/2$ are the effective Young's modulus and effective radius respectively. By replacing the coordinate $x_i$ by the displacement $u_i$ of the $i$th buckyball from its equilibrium position in the chain, the equation of motion for each buckyball may be further written as $\ddot{u}_i^{\frac{3}{2}} = (u_{i-1} - u_i)^{\frac{3}{2}} - (u_i - u_{i+1})^{\frac{3}{2}}$ **(2)** This is widely used for granular materials.

### *MD simulation of one-dimensional $C_{60}$ chain.*

The forces on both impactor and receiver plates are normalized as $FR_{C_{60}}/Eh^3$, and the representative impact force attenuation for 100 $C_{60}$ particles is shown in Figure 2 (where the positive value stands for compression force along the impact velocity direction). A sharp and narrow impact pulse is initiated once the top plate collides with the buckyball system and it drops to nearly zero at about 0.02 ns ($\Delta\tau_1 \approx 0.02\,\text{ns}$), indicating that the compressive stress wave is traveling towards the receiver. The receiver does not experience any force until the stress wave arrives at $t = t_1$ (shown in Figure 2); from which the average traveling speed of the stress wave is estimated as $u_0 = L/t_1 \approx 1252\,\text{m/s}$ and thus the system equivalent modulus is $E = u_0^2 \rho_{C_{60}} \approx 3.10\,\text{GPa}$ for the specific impact loading condition (impact energy of 6.49 eV and impact speed of 500 m/s). Once the stress wave reaches the receiver, it reflects back and if it successfully travels back to the impactor, a secondary impact impulse would form at $t = t_2$ (shown in Figure 2) and thus causes the speed of the ricochet impactor increase again. The peak transmitted force on the receiver is about 42.27% of the original peak force on the impactor, after force attention of 100 $C_{60}$ buckyballs. About 93.75% of the impactor kinetic energy (i.e. impact energy) is dissipated by the system, therefore, one may define the energy mitigation rate as $\eta = 0.9375$. The effect of buckyball number on the energy mitigation rate is discussed later.

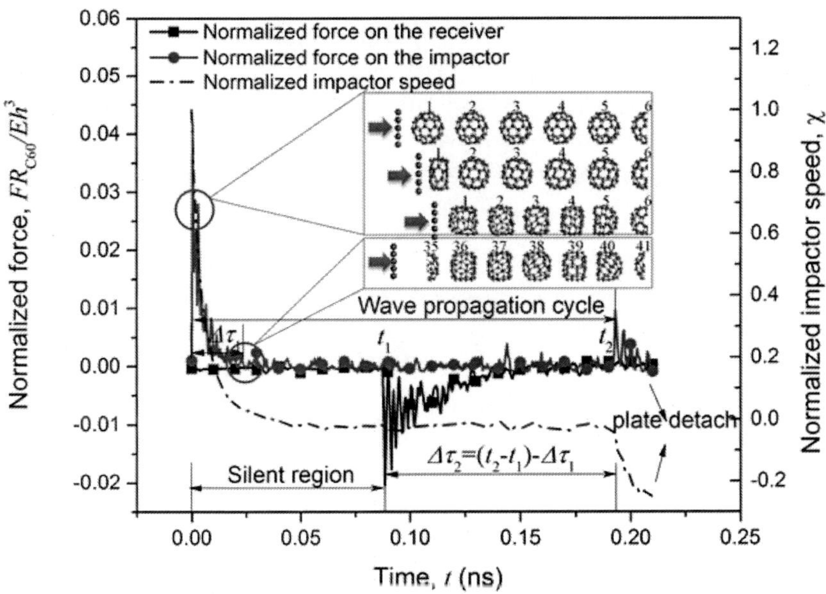


Figure 2. Normalized force time history and impactor velocity history of $C_{60}$ chain containing 100 buckyballs, when the impact energy is 6.49 eV and impact speed is 500 m/s.

doi:10.1371/journal.pone.0064697.g002

According to the force equilibrium and mass continuity, the stress of a particular mass point during the wave propagation is $\sigma = v\sqrt{E\rho_{C_{60}}}$ [42], and the relation between stress $\sigma_b$ and $\sigma_r$(at the interface of buckyball and the rigid plate respectively) may be expressed as [42]$\frac{\sigma_r}{\sigma_b}=\frac{\rho_r c_r-\rho_b c_b}{\rho_r c_r+\rho_b c_b}$**(3)** where $\rho_r$ and $\rho_b$ are densities and $c_b$ and $c_r$ are wave speeds of the material on the two sides of the interface respectively. Similarly, the stress wave speed may be written as [42]$\frac{v_r}{v_b}=\frac{\rho_b c_b-\rho_r c_r}{\rho_b c_b+\rho_r c_r}$**(4)** Since the receiver is fixed as a rigid body in this study, $\rho_r c_r=\infty$, such that $\sigma_r/\sigma_b=1$ and $v_r/v_b=-1$ which means that the stress wave propagates back to impactor at the same speed. After the reflective wave travels through 100 $C_{60}$ buckyballs, the magnitude of force on impactor reduces to 21.95% of the original force. On the other hand, the transmitted force pulse duration is about 5.4 times of that on the impactor, i.e. $\Delta\tau_1/\Delta\tau_2\approx 0.185$, showing a prominent stress wave mitigation effect. The major energy mitigation effect results from the stress wave attenuation caused by the reflections among buckyball walls, similar as that found in previous research in granular system [22], [25], [29], [30], [43], as well as the van der Waals interactions between buckled layers and similar energy absorption mechanism revealed in carbon nanotubes in Ref. [17], [18], [44]. In addition, about 1.5% of the impact energy may be converted to the kinetic energy of the atoms within $C_{60}$.

### *Dissipative Hertzian model*

As the method adopted in Ref [45] to include the dissipation term to Eq. (2), from MD simulation, the following relationship can be fitted: $\ddot{\delta}_i=A\left[(\delta_i)^{\frac{3}{2}}-(\delta_{i+1})^{\frac{3}{2}}\right]+\alpha(\delta_i-\delta_{i+1})^6$ **(5)**
where $A=E/3(1-\nu^2)m_{\text{buckyball}}\cdot\sqrt{2R_{C_{60}}}$, the second term implies dissipation which is fitted based on the force-displacement curve at large deformation in our previous study [20],[38] and its coefficient $\alpha=61.32\ \mathrm{m}^{-5}\cdot\mathrm{s}^{-2}$. This relationship is valid for systems with large number of $C_{60}$ buckyballs at all loading conditions as long as the Hertzian contact law holds.Figure 3 shows the maximum force on the $i$th ball, $F_{i,\max}(t)$, of the dissipative model (Eq. (5)), which is consistent with the MD results of $C_{60}$ chains.

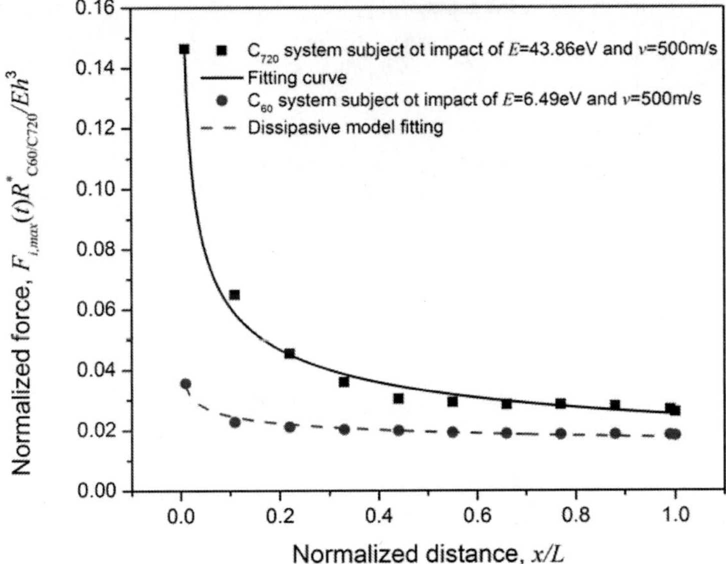


Figure 3. The normalized force distribution on selected buckyballs in $C_{60}$ and $C_{720}$ chain systems.

doi:10.1371/journal.pone.0064697.g003

## Dynamic response of $C_{720}$ chain system

The large non-recoverable deformation of $C_{720}$ makes the Hertzian contact law invalid. The energy mitigation behaviors are investigated using MD simulations. Typical normalized force history curves of the impactor and receiver are shown in Figure 4, where 100 $C_{720}$ are studied. In terms of stress wave traveling, its average speed is $u_0 = L/t_1 \approx 509.8\,\mathrm{m/s}$, and thus the system equivalent Young's modulus is $E = u_0^2 \rho_{C_{720}} \approx 0.536\,\mathrm{GPa}$, which means the $C_{720}$chain system is more "compliant" than $C_{60}$. In our previous work, the "non-recovery" phenomenon is proven to be only strain determined, regardless of the impact mass and velocity[38]. During preliminary simulations, we also confirm that the "non-recovery" phenomenon in $C_{540}$ is impact-condition independent.

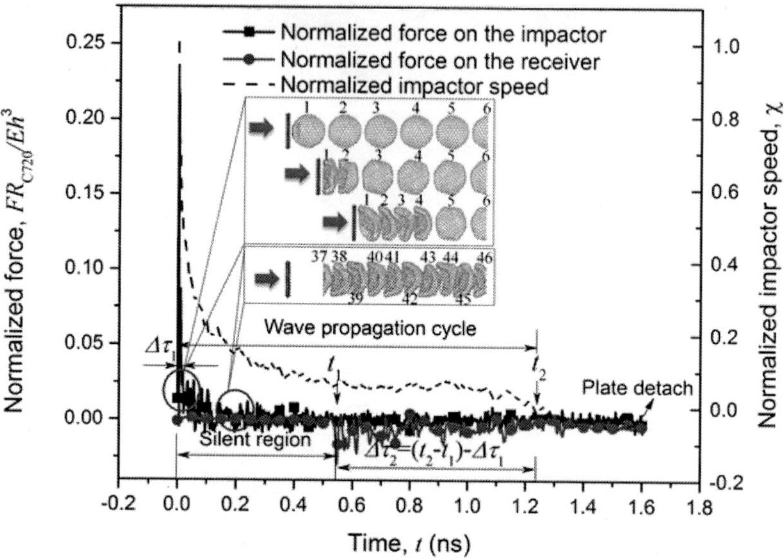


Figure 4. Normalized force history and impactor velocity history of $C_{720}$ chain containing 100 buckyballs, when the impact energy is 6.49 eV and impact speed is 500 m/s.

doi:10.1371/journal.pone.0064697.g004

From energy mitigation perspective, a very sharp initial impulse is attenuated to a much milder and longer impulse. The ratio of the peak magnitude and duration between the original and transmitted impulses are $F_r/F_i \approx 13.2\%$ (where the subscript $r$ and $i$ refer to the receiver and impactor respectively) and $\Delta\tau_1/\Delta\tau_2 \approx 0.0290$. The force reduction and duration elongation are much higher than that in $C_{60}$ chain system due to the buckled-through shape of $C_{720}$ during impact. Therefore, van der Waals interactions between buckled and "stickered" layers may contribute more energy dissipation compared to its counterpart in $C_{60}$ system due to the un-recoverable deformation. Also, with the buckled morphorlogy of $C_{720}$, the covalent potential energy also increase via the consumption of external impact energy. Moreover, about 12% of the impact energy could be mitigated in the form of atom kinetic energy which also contributes the superiority of energy dissipation for $C_{720}$ system. The power-law-like dissipative model for contact force attenuation $F_{i,\max}(t)$ at various buckyballs still applies (see Fig. 3), indicating a fast force decay along the wave propagation direction. In the meantime, over 99% of the impact energy is mitigated to the kinetic energy and strain energy of buckyballs.

## Parametric Study and Discussions

A parametric study is carried out where the impact speed is varied from $v_0 = 100$ m/s to 1000 m/s, and the impact mass per carbon atom is varied at $\varphi = m_{impactor}/m_{buckyball}$ = 1.73 to 13.87 for both the $C_{60}$ and $C_{720}$ chains containing 100 buckyballs. The initial impact speed is normalized as $\upsilon = v_0/u_0$ (where $u_0 = 1252\,m/s$ and $u_0 = 509.8\,m/s$ are used for $C_{60}$ and $C_{720}$ chains respectively); the stress wave propagation speed (calculated based on the time when the wave transmits through the chain, which is dependent on the number of buckyballs) is normalized as $\mu = u/u_0$. The corresponding fitting curves of the suggested models are also shown in Figures 5 and 6.

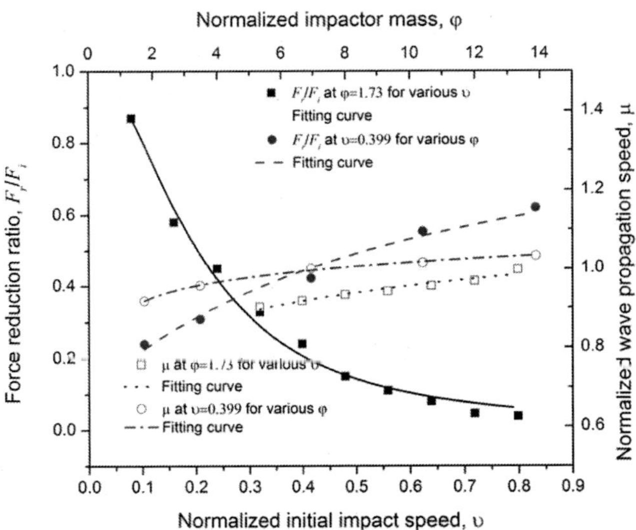


Figure 5. Force reduction ratio $F_r/F_i$ and normalized wave propagation speed $\mu$ under various impact speeds ($0.080 \leq \upsilon \leq 0.80$) with fixed impact mass ($\varphi = 1.73$), as well as various impact masses ($1.73 \leq \varphi \leq 13.87$) with fixed impact speed ($\upsilon = 0.40$) for $C_{60}$ chain system containing 100 buckyballs.

Nonlinear models are suggested to fit the computational data.

doi:10.1371/journal.pone.0064697.g005

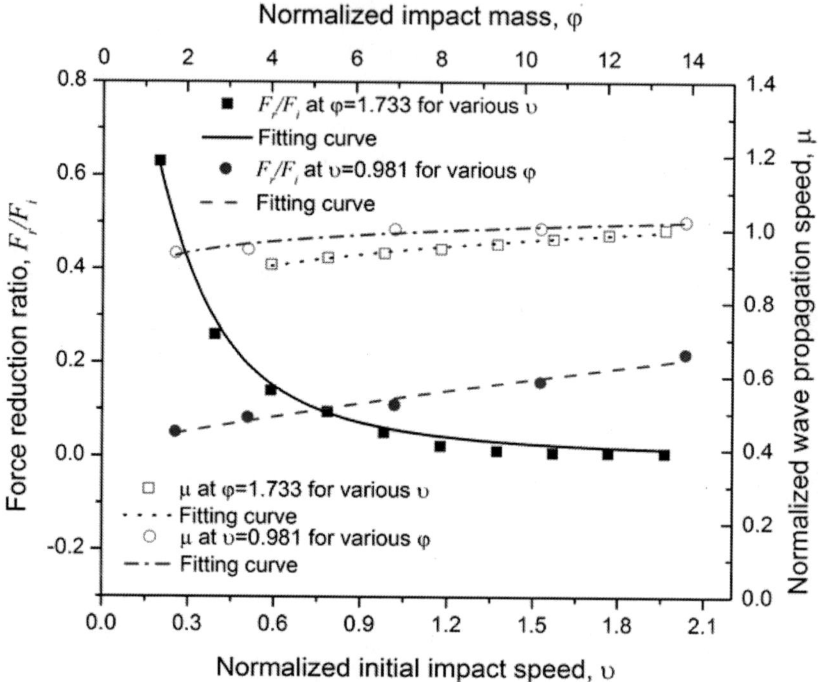


Figure 6. Force reduction ratio $F_r/F_i$ and normalized wave propagation speed $\mu$ under various impact speeds ($0.196 \leq \upsilon \leq 1.96$) with fixed impact mass ($\varphi = 1.73$), as well as various impact masses ($1.73 \leq \varphi \leq 13.87$) with fixed impact speed ($\upsilon = 0.98$) for $C_{720}$ chain system containing 100 buckyballs.

Nonlinear models are suggested to fit the computational data.

doi:10.1371/journal.pone.0064697.g006

Effects of initial impact speed and mass on $C_{60}$ chain

### ***Force attenuation***

The force reduction ratio $F_r/F_i$ and normalized wave propagation speed $\mu$ are two indices employed to evaluate the mitigation properties, shown in Figure 5. Following Reid and Peng[46], the enhanced dynamic stress $\sigma^*$ can be expressed as $\sigma^* = \sigma_{cr} + \frac{\rho_{buckyball} v^2}{\varepsilon_D}$ **(6)** where $\sigma_{cr}$ is the crushing strength of buckyball and the $\varepsilon_D$ is the material strain attained behind the wave front. $v$ is the particle velocity at a certain time *t*. By keeping the impact mass constant, the particle velocity, $v \propto v_0$ [46]. Assuming the contact area keeps a constant as $A_0$, one may come to $F_r/F_i = \sigma_r^*/\sigma_i^*$. Thus, the force reduction ratio under a fixed impact mass (but different impact velocities) is

$$\frac{F_r}{F_i} = \frac{\sigma_{cr}}{\sigma_{cr} + k\frac{\rho_{buckyball} v_0^2}{\varepsilon_D}} = \frac{1}{1 + k\frac{\rho_{buckyball} v_0^2}{\sigma_{cr}\varepsilon_D}} \quad \textbf{(7)}$$

where $k$ is the linear coefficient between $v$ and $v_0$. Alternatively, under this condition, we may fit the equation in the form of $\frac{F_r}{F_i} = \frac{1}{1+av^2}$ **(8)** where $a$ is material-related parameter and from Figure 5 it yields $a \approx 24.05$ for the present system.

One may also regard the buckyball system as a non-linear spring damping system whose stiffness is only slightly affected by the mass of the impactor. Such a damping system reduces the force in the receiver by extending the functioning time over a longer time period. When the impact speed remains the same but impact mass is different, the following form is fitted to describe the force attenuation

$$\frac{F_r}{F_i} = 1 \Big/ \left(1 + \beta \ln\left(1 + \frac{\gamma}{\varphi + 1}\right)\right) \quad \textbf{(9)}$$

where $\beta = 39.6$ and $\gamma = 0.247$ for the present system.

Eqs. (8) and (9) in together reveal that the one-dimensional $C_{60}$ chain system has a better mitigation performance under the condition of higher impact speed with smaller mass, in terms of the force attenuation to alleviate the transmitted load on objects to be protected.

### ***System equivalent Young's modulus***

The system equivalent Young's modulus may be characterized via the elongation of wave propagation speed. The mitigation behavior is still dominated by impact energy, which means that changing the impactor mass or speed may vary the mitigation performance. The ratio between dynamic stress $\sigma_{dynamic}$ and static stress $\sigma_{static}$ for rate-sensitive material may be expressed as [47]

$$\frac{\sigma_{dynamic}}{\sigma_{static}} = 1 + \left(\frac{\dot{\varepsilon}}{D}\right)^{1/q} \quad \textbf{(10)}$$

where $\dot{\varepsilon}$ is the strain rate, $D$ and $q$ are constants for a particular material. With the relation between stress and Young's modulus as well as the strain rate and velocity, one may fit the normalized wave propagation speed with varying impact speed $\mu$ (yet same impact mass) in the form of $\mu = \left(\frac{v}{D}\right)^{1/q}$ **(11)** where $D = 0.937$ and $q = 9.835$. Combining the two equations above yields the relationship under various impact masses (but same impact speed):

$$\mu = \left(\frac{\sqrt{A\varphi^2 - 1}}{B}\right)^{1/p} \quad \textbf{(12)}$$

where $A = 1.41$, $B = 8.97$ and $p = 18.8$ through the best fitting.

Note that when the number of buckyballs in the system increases, the effective system rigidity becomes smaller due to the longer stress wave transmission. Therefore, the fitted formula for calculating the stress wave speed and the corresponding equivalent rigidity are only valid for this specific system under subscribed loading conditions. However, these parametric values may become numerically convergent under certain impact mass once the number of buckyball reaches the threshold value which is discussed later.

Effect of initial impact speed and mass on $C_{720}$ chain

## *Force attenuation*

To evaluate the energy mitigation performance of $C_{720}$ chain, the force reduction ratio $F_r/F_i$ and normalized wave propagation speed $\mu$ are employed in Figure 6. The $F_r/F_i$ value reduces sharply in the relatively low impact speed domain and becomes stable once the impact speed exceeds over 500 m/s. Eq. (8) still applies with $a \approx 15.77$ via best fitting.

Due to the non-recoverable deformation of $C_{720}$, the impact mass poses stronger influence over the force reduction ratio because the larger mass makes the first few buckyballs easier to buckle. With the impactor mass increasing, the force reduction is also more prominent than that in $C_{60}$ chain system. Similarly, Eq. (10) may be applied with $\beta=119$ and $\gamma=0.486$.

## System equivalent Young's modulus

Similarly, we may also take the form of Eqs. (11) and (12) to describe the system equivalent rigidity based on stress wave propagation speed. With the impact speed increases, the average wave propagation speed also increases, leading to a much stiffer system in terms of rigidity. InEq. (11), the fitting parameters are $D=2.12$ and $q=12.0$ for the $C_{720}$ buckyball system. By taking the derivative of Eq. (11), the variation rate in $C_{60}$ is more prominent than that of $C_{720}$, indicating that $C_{60}$ exhibit even higher effective stiffness than $C_{720}$ under very high impact speed situations. The fitting of Eq. (12) yields $A=1.20$, $B=9.49$ and $p=23.3$ for the $C_{720}$chain system, indicating that the effect of impactor mass is less on $C_{720}$ than that on $C_{60}$chain. Again, these fitted equations are only valid for the protective system with particular number of buckyball under the specific loading conditions. System rigidity would also alter accordingly if any of the corresponding factors change.

### Effect of buckyball size

The ratio between the initial and transmitted impulse duration, $\Delta\tau_1/\Delta\tau_2$, is also an important indicator for energy mitigation. The buckling forces for larger size buckyballs are smaller, owing to the buckling phenomenon. Figure 7 shows the relation between $\Delta\tau_1/\Delta\tau_2$ and normalized buckyball diameter $\Omega=R_{\text{buckyball}}/R_{C_{60}}$ at the impact speed of 500 m/s with the same impactor mass per carbon atom. The sizes of all buckyball involved here are labeled in Figure 7. The $\Delta\tau_1/\Delta\tau_2$ values decay in a power-law manner as the buckyball size increases. More importantly, a sudden drop is observed between $C_{320}$ and $C_{540}$ where the non-recovery phenomenon starts to appear. Once the buckyballs stay in a buckled morphology, the layered and densified structure would create more barriers to transmit the stress waves and the waves are attenuated through the wave reflection among interfaces of buckled shapes. The numerical results may be

fitted as
$$\begin{cases} \Delta\tau_1/\Delta\tau_2 = 0.184\Omega^{-0.576}, & 1 \leq \Omega \leq 2.20 \\ \Delta\tau_1/\Delta\tau_2 = 0.0660 - 0.0115\Omega, & 2.89 \leq \Omega \leq 3.22 \end{cases} \quad (13)$$

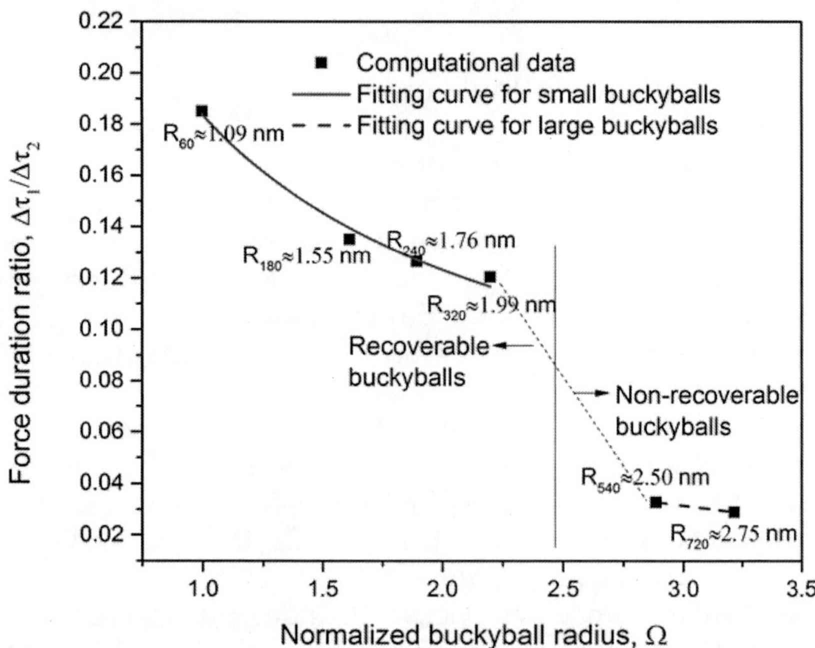


Figure 7. The impulse duration ratios $\Delta\tau_1/\Delta\tau_2$ between the impactor and receiver for various buckyballs including $C_{60}$, $C_{180}$, $C_{240}$, $C_{320}$, $C_{540}$ and $C_{720}$ by normalized buckyball radii $\Omega$ at the impact speed of 500 m/s with the same $\varphi$ value in each buckyball.

doi:10.1371/journal.pone.0064697.g007

## Effect of buckyball number

As aforementioned, with the change of buckyball number within the protective system, stress wave propagation characteristics as well as the equivalent system rigidity alters, which may influence the energy mitigation ability of the system for both $C_{60}$ and $C_{720}$ systems. The energy mitigation rate $\eta$ is calculated for systems with buckyball numbers varying from 1 to 200 under the specific impact condition. In Figure 8, one may clearly observe that nonlinear increase on $\eta$ with the buckyball number for both systems. The increasing rate becomes much milder in longer buckyball chains, indicating that there may be a certain length threshold beyond which the system acquires high-efficiency impact wave mitigation. In addition, to reach the same mitigation ability, fewer buckyballs are needed to for larger particles; for example, the system with about 20 $C_{720}$ buckyballs may mitigate over 99% of the impactor kinetic energy (i.e. $\eta > 99\%$ ), whereas it would take about 80 $C_{60}$ buckyballs to reach $\eta > 90\%$, showing another superiority of $C_{720}$ system without the system mass and volume constrain in application.

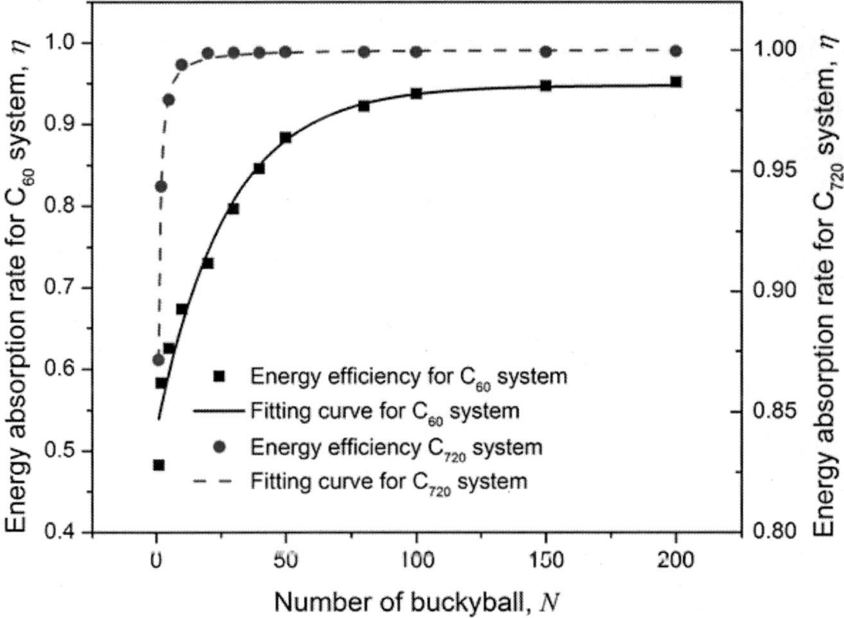


Figure 8. Relations between energy absorption rate and buckyball number for both $C_{60}$ and $C_{720}$ systems at the impact speed $\upsilon = 0.40$ for $C_{60}$ chain system and$\upsilon = 0.98$ for $C_{720}$ chain system, and the impact mass of $\varphi = 1.73$ for both systems.

doi:10.1371/journal.pone.0064697.g008

From systematic simulations, one may also summarize an empirical law at the impact speed $\upsilon=0.40$ for $C_{60}$ chain system and $\upsilon=0.98$ for $C_{720}$ chain system, and the impact mass of $\varphi=1.73$ for both systems to describe the relation between buckyball number $N$ ($N$>0) and $\eta$ as

$$\begin{cases} \eta=0.9519-0.4872\cdot0.9628^{N}, & \text{for } C_{60} \text{ system} \\ \eta=0.8711+0.1292(1-N^{-1.182}), & \text{for } C_{720} \text{ system} \end{cases} \quad (14)$$

and Eq. (14) may serve as a guidance for engineering design.

## Concluding Remarks

In this paper, the impact mitigation characteristics of a long one dimensional buckyball chain are investigated, which can be extended to granular buckyballs of simple cubic packing. Representative small and large buckyballs, i.e. $C_{60}$ and $C_{720}$ under high speed impact loadings are studied. The impact energy, size and number of buckyballs, are varied in a systematic manner. With relatively small elastic deformations of $C_{60}$ buckyballs during impact, a mechanical model based on Hertz contact law is proposed, with critical parameters calibrated via MD simulations for given impact loading conditions. Energy mitigation is illustrated through force impulse history difference between the impact and receiver. The stress wave propagation speed, the reduction of peak impulse force, and the impulse duration ratio are studied to reveal the dynamic response of the system. The major energy dissipation mechanism for the buckyball chain is the wave reflection among the deformation layers, covalent potential energy, van der Waals interactions as well as the atomistic kinetic energy. These terms may have higher contribution to energy dissipation in $C_{720}$ system with non-recoverable morphologies. Moreover, Buckyball systems are investigated under various impact speeds and impact masses. The smaller mass and higher impact speed results in a higher impulse force attenuation effect, as well as higher system stiffness and shorter wave propagation time. Over 99% and 90% of impact energy for $C_{720}$ and $C_{60}$ chain systems could be mitigated under particular impact conditions respectively and thus a promising buckyball based stress wave mitigation system is suggested. The results may shed lights on the research and development of novel impact/blast protection system.

## Author Contributions

Conceived and designed the experiments: JX YX XC. Performed the experiments: JX YL. Analyzed the data: JX XC. Contributed reagents/materials/analysis tools: JX YX XC. Wrote the paper: JX XC.

## REFERENCES


1. Mahfuz H, Zhu YH, Haque A, Abutalib A, Vaidya U, et al. (2000) Investigation of high-velocity impact on integral armor using finite element method. International Journal of Impact Engineering 24: 203–217. doi: 10.1016/s0734-743x(99)00047-0
2. Deshpande VS, Fleck NA (2000) High strain rate compressive behaviour of aluminium alloy foams. International Journal of Impact Engineering 24: 277–298. doi: 10.1016/s0734-743x(99)00047-0
3. Jones N (2010) Energy-absorbing effectiveness factor. International Journal of Impact Engineering 37: 754–765. doi: 10.1016/s0734-743x(99)00047-0
4. Lopatnikov SL, Gama BA, Haque MJ, Krauthauser C, Gillespie JW (2004) High-velocity plate impact of metal foams. International Journal of Impact Engineering 30: 421–445. doi: 10.1016/s0734-743x(99)00047-0
5. Nilakantan G, Gillespie JW (2012) Ballistic impact modeling of woven fabrics considering yarn strength, friction, projectile impact location, and fabric boundary condition effects. Composite Structures 94: 3624–3634. doi: 10.1016/s0734-743x(99)00047-0
6. Ha-Minh C, Boussu F, Kanit T, Crepin D, Imad A (2011) Analysis on failure mechanisms of an interlock woven fabric under ballistic impact. Engineering Failure Analysis 18: 2179–2187. doi: 10.1016/j.engfailanal.2011.07.011
7. Hou YQ, Sun BZ, Gu BH (2011) An analytical model for the ballistic impact of three dimensional angle-interlock woven fabric penetrated by a rigid cylindro-spherical projectile. Textile Research Journal 81: 1287–1303. doi: 10.1016/j.engfailanal.2011.07.011
8. Cuong HM, Kanit T, Boussu F, Imad A (2011) Numerical multi-scale modeling for textile woven fabric against ballistic impact. Computational Materials Science 50: 2172–2184. doi: 10.1016/j.engfailanal.2011.07.011
9. Tan CY, Akil HM (2012) Impact response of fiber metal laminate sandwich composite structure with polypropylene honeycomb core. Composites Part B-Engineering 43: 1433–1438. doi: 10.1016/j.engfailanal.2011.07.011
10. Xia F, Wu XQ (2010) Work on Impact Properties of Foam Sandwich Composites with Different Structure. Journal of Sandwich Structures & Materials 12: 47–62. doi: 10.1016/j.engfailanal.2011.07.011
11. Xu J, Li YB, Chen X, Ge DY, Liu BH, et al. (2011) Automotive windshield - pedestrian head impact: Energy absorption capability of interlayer material.

International Journal of Automotive Technology 12: 687–695. doi: 10.1007/s12239-011-0080-2

12. Qin QH, Wang TJ (2011) Low-velocity heavy-mass impact response of slender metal foam core sandwich beam. Composite Structures 93: 1526–1537. doi: 10.1007/s12239-011-0080-2
13. Vaidya UK, Pillay S, Bartus S, Ulven CA, Grow DT, et al. (2006) Impact and post-impact vibration response of protective metal foam composite sandwich plates. Materials Science and Engineering a-Structural Materials Properties Microstructure and Processing 428: 59–66. doi: 10.1007/s12239-011-0080-2
14. Lopatnikov SL, Gama BA, Haque MJ, Krauthauser C, Gillespie JW, et al. (2003) Dynamics of metal foam deformation during Taylor cylinder-Hopkinson bar impact experiment. Composite Structures 61: 61–71. doi: 10.1007/s12239-011-0080-2
15. Mylvaganam K, Zhang LC (2006) Energy absorption capacity of carbon nanotubes under ballistic impact. Applied Physics Letters 89. doi: 10.1007/s12239-011-0080-2
16. Pan H, Zhang YW, Shenoy VB, Gao HJ (2011) Ab Initio Study on a Novel Photocatalyst: Functionalized Graphitic Carbon Nitride Nanotube. Acs Catalysis 1: 99–104. doi: 10.1136/jcp.40.4.474-d
17. Yang XD, He PF, Gao HJ (2012) Competing elastic and adhesive interactions govern deformation behaviors of aligned carbon nanotube arrays. Applied Physics Letters 101. doi: 10.1136/jcp.40.4.474-d
18. Yang XD, He PF, Gao HJ (2011) Modeling frequency- and temperature-invariant dissipative behaviors of randomly entangled carbon nanotube networks under cyclic loading. Nano Research 4: 1191–1198. doi: 10.1136/jcp.40.4.474-d
19. Xu BX, Liu L, Zhou QL, Qiao Y, Xu J, et al. (2011) Energy Dissipation of Nanoporous MFI Zeolite Under Dynamic Crushing. Journal of Computational and Theoretical Nanoscience 8: 881–886. doi: 10.1136/jcp.40.4.474-d
20. Xu J, Sun Y, Li Y, Xiang Y, Chen X (2013) Molecular dynamics simulation of impact response of buckyballs. Mechanics Research Communications 49: 8–12. doi: 10.1136/jcp.40.4.474-d
21. Lu W, Punyamurtula VK, Qiao Y (2011) An energy absorption system based on carbon nanotubes and nonaqueous liquid. International Journal of Materials Research 102: 587–590. doi: 10.3139/146.110377
22. Grujicic M, Pandurangan B, Bell WC, Bagheri S (2012) Shock-Wave Attenuation and Energy-Dissipation Potential of Granular Materials. Journal of Materials Engineering and Performance 21: 167–179. doi: 10.3139/146.110377
23. Park J, Palumbo DL (2009) Damping of Structural Vibration Using Lightweight Granular Materials. Experimental Mechanics 49: 697–705. doi: 10.3139/146.110377
24. Nesterenko VF (2001) Dynamics of heterogeneous materials. New Yor, NY, USA: Springer.
25. Manjunath M, Awasthi AP, Geubelle PH (2012) Wave propagation in random

granular chains. Physical Review E 85. doi: 10.3139/146.110377

26. Antypov D, Elliott JA, Hancock BC (2011) Effect of particle size on energy dissipation in viscoelastic granular collisions. Physical Review E 84. doi: 10.1103/physreve.84.021303
27. Spannuth MJ, Mueggenburg NW, Jaeger HM, Nagel SR (2004) Stress transmission through three-dimensional granular crystals with stacking faults. Granular Matter 6: 215–219. doi: 10.1103/physreve.84.021303
28. Mueggenburg NW, Jaeger HM, Nagel SR (2002) Stress transmission through three-dimensional ordered granular arrays. Physical Review E 66. doi: 10.1103/physreve.84.021303
29. Kruyt NP, Agnolin I, Luding S, Rothenburg L (2010) Micromechanical study of elastic moduli of loose granular materials. Journal of the Mechanics and Physics of Solids 58: 1286–1301. doi: 10.1103/physreve.84.021303
30. Carretero-Gonzalez R, Khatri D, Porter MA, Kevrekidis PG, Daraio C (2009) Dissipative Solitary Waves in Granular Crystals. Physical Review Letters 102. doi: 10.1103/physreve.84.021303
31. Sen S, Hong J, Bang J, Avalos E, Doney R (2008) Solitary waves in the granular chain. Physics Reports-Review Section of Physics Letters 462: 21–66. doi: 10.1016/j.physrep.2007.10.007
32. Gui XC, Wei JQ, Wang KL, Cao AY, Zhu HW, et al. (2010) Carbon Nanotube Sponges. Advanced Materials 22: 617–+. doi: 10.1002/adma.200902986
33. Wang CM, Zhang YY, Xiang Y, Reddy JN (2010) Recent Studies on Buckling of Carbon Nanotubes. Applied Mechanics Reviews 63. doi: 10.1016/j.physrep.2007.10.007
34. Chandraseker K, Mukherjee S (2007) Atomistic-continuum and ab initio estimation of the elastic moduli of single-walled carbon nanotubes. Computational Materials Science 40: 147–158. doi: 10.1016/j.physrep.2007.10.007
35. Yakobson BI, Brabec CJ, Bernholc J (1996) Nanomechanics of carbon tubes: Instabilities beyond linear response. Physical Review Letters 76: 2511–2514. doi: 10.1016/j.physrep.2007.10.007
36. Man ZY, Pan ZY, Ho YK (1995) The rebounding of C60 on graphite surface: a molecular dynamics simulation. Physics Letters A 209: 53–56. doi: 10.1016/0375-9601(95)00768-7
37. Pan ZY, Man ZY, Ho YK, Xie J, Yue Y (1998) Energy dependence of C60–graphite surface collisions. Journal of Applied Physics 83: 4963–4967. doi: 10.1016/0375-9601(95)00768-7
38. Xu J, Li Y, Xiang Y, Chen X (2013) Investigation of Energy Absorption Ability of Buckyball $C_{720}$ at Low Impact Speed Based on Molecular Dynamic Simulation. Nanoscale Research Letters 8: 54. doi: 10.1186/1556-276X-8-54
39. Plimpton S (1995) Fast parallel algorithms for short-range molecular dynamics. Journal of Computational Physics 117: 1–19. doi: 10.1016/0375-9601(95)00768-7

40. Girifalco LA, Hodak M (2002) Van der Waals binding energies in graphitic structures. Physical Review B 65. doi: 10.1016/0375-9601(95)00768-7
41. Johnson KL (1985) Contact Mechanics. Cambridge, England: Cambridge University Press.
42. Lu G, Yu TX (2003) Energy absorption of structures and materials: Woodhead Publishing Ltd.
43. Daraio C, Nesterenko VF, Herbold EB, Jin S (2006) Energy trapping and shock disintegration in a composite granular medium. Physical Review Letters 96. doi: 10.1103/physrevlett.96.058002
44. Fraternali F, Blesgen T, Amendola A, Daraio C (2011) Multiscale mass-spring models of carbon nanotube foams. Journal of the Mechanics and Physics of Solids 59: 89–102. doi: 10.1103/physrevlett.96.058002
45. Herbold EB, Nesterenko VF (2007) Solitary and shock waves in discrete strongly nonlinear double power-law materials. Applied Physics Letters 90. doi: 10.1103/physrevlett.96.058002
46. Reid SR, Peng C (1997) Dynamic uniaxial crushing of wood. International Journal of Impact Engineering 19: 531–570. doi: 10.1016/s0734-743x(97)00016-x
47. Cowper GR, Symonds PS (1957) Strain hardening and strain-rate effects in the impact loading of cantilever beams. Brown University Division of Applied Mathematics doi: 10.1016/s0734-743x(97)00016-x

# Chapter 7

# THE NANOSTRUCTURE OF MYOENDOTHELIAL JUNCTIONS CONTRIBUTES TO SIGNAL RECTIFICATION BETWEEN ENDOTHELIAL AND VASCULAR SMOOTH MUSCLE CELLS

Jens Christian Brasen*, Jens Christian Brings Jacobsen, Niels-Henrik Holstein-Rathlou

Department of Biomedical Sciences, University of Copenhagen, Copenhagen, Denmark

## ABSTRACT

Micro-anatomical structures in tissues have potential physiological effects. In arteries and arterioles smooth muscle cells and endothelial cells are separated by the internal elastic lamina, but the two cell layers often make contact through micro protrusions called myoendothelial junctions. Cross talk between the two cell layers is important in regulating blood pressure and flow. We have used a spatiotemporal mathematical model to investigate how the myoendothelial junctions affect the information flow between the two cell layers. The geometry of the model mimics the structure of the two cell types and the myoendothelial junction. The model is implemented as a 2D axi-symmetrical model and solved using the finite element method. We have simulated diffusion of $Ca^{2+}$ and $IP_3$ between the two cell types and we show that the micro-anatomical structure of the myoendothelial junction in itself may rectify a signal between the two cell layers. The rectification is caused by the asymmetrical structure of the myoendothelial junction. Because the head of the myoendothelial junction is separated from the cell it is attached to by a narrow neck region, a signal generated in the neighboring cell can easily drive a concentration change in the head of the myoendothelial protrusion. Subsequently the signal can be amplified in the head, and activate the entire cell. In contrast, a signal in the cell from which the myoendothelial junction originates will be attenuated and delayed in the neck region as it travels into the head of the myoendothelial junction and the neighboring cell.

## INTRODUCTION

Information processing in tissues often relies on unidirectional flow of information. Such unidirectional flow is found in e.g. synapses of the nervous system [1]. Similar specialized anatomical structures that potentially enable signal rectification are also found in arteries and arterioles. Such vessels consist of a single layer of endothelial cells (ECs, see Table 1 for a full list of abbreviations), which lines the lumen, surrounded by one or more layers of smooth muscle cells (SMCs). The two cell types are separated by the internal elastic lamina [2], [3]. However, ECs and SMCs make occasional contacts through myoendothelial (i.e. muscle-endothelial) junctions (MEJs), which are mushroom shaped protrusions that project from one cell layer and traverse the internal elastic lamina to make contact with the other layer [2]–[5]. The MEJ can extend from either cell layer depending on the tissue and organism [2]–[4]. Gap junctions in the MEJ connect the cytoplasm of the two cells and are important in myoendothelial signal transduction [6], [7]. The gap junction itself is permeable to ions and small molecules (<1 kDa) including $Ca^{2+}$ and $IP_3$ [8]–[11].

Table 1. List of abbreviations.

doi:10.1371/journal.pone.0033632.t001

| Abbreviation | Name |
|---|---|
| EC | Endothelial cell |
| SMC | Smooth muscle cell |
| MEJ | Myoendothelial junction |
| ER | Endoplasmic reticulum |
| $IP_3$ | Inositol 1,4,5-trisphosphate |
| IP3R | Inositol 1,4,5-trisphosphate receptor (a $Ca^{2+}$ channel) |

doi:10.1371/journal.pone.0033632.t001

The functional properties of the MEJ have been studied in mouse and rat mesenteric arteries where they extend from the EC layer to contact the SMC layer [6], [7]. A $Ca^{2+}$ signal initiated in the SMC layer (e.g. by stimulation with phenylephrine) is apparently transmitted to the ECs by diffusion of $IP_3$, whereas propagation of the signal in the opposite direction does not take place [12], . The directionality has been proposed to be due to rectification in heterodimeric gap junctions or to differences in distribution of $IP_3$-receptor channels and other proteins [14]. However, the micro-anatomical structure of the MEJ has never been considered in this regard, and here we show that the structure of the MEJ alone can rectify a signal, which is mediated by a diffusible molecule between the two cell layers. The model shows that in an arteriole, the structure of the MEJ enables a diffusible molecule to propagate from the SMC into the MEJ from where it can spread into the body of the EC. In contrast, signaling in the opposite direction is less likely to happen due to the structure of the MEJ.

# MATERIALS AND METHODS

## Description of the Models

To address the apparent asymmetry in myoendothelial signaling we constructed a spatiotemporal model of the EC-SMC contact (Fig. 1A–B). The structure is based on electron microscopic images of small rat and mouse mesenteric arteries [5], [15]. We assumed radial symmetry around the long axis of the MEJ (Fig. 1A) which enabled us to implement the model as a 2D axi-symmetrical model (Fig. 1B–C).

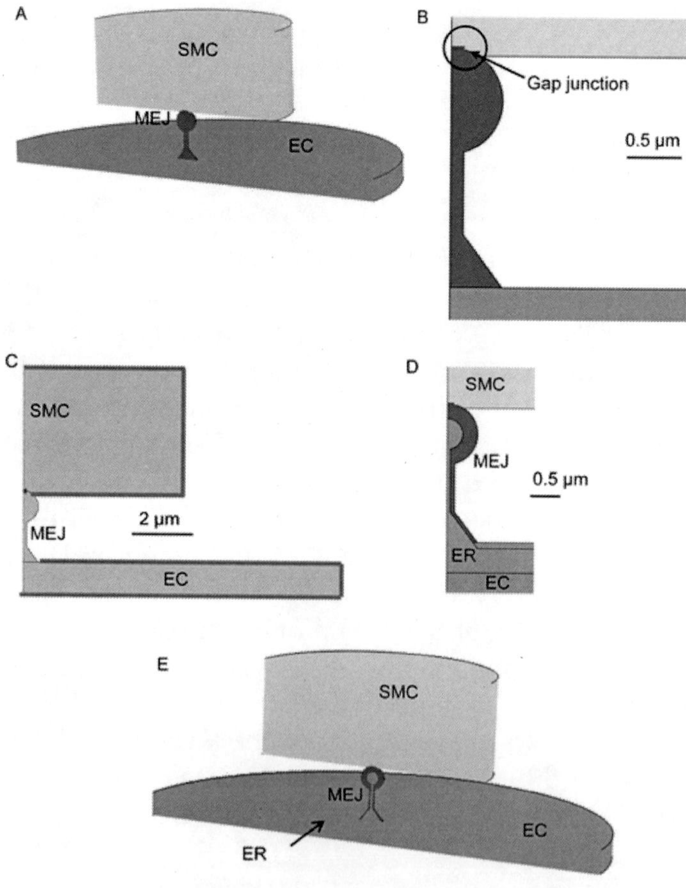


Figure 1. Description of the models.

A, B) Model of the interaction between a smooth muscle cell (SMC) and an endothelial cell (EC) through the myoendothelial junction (MEJ). The SMC is shown in yellow, the EC in green and the MEJ in red. The SMC and MEJ are coupled in the area highlighted with the black circle at the blue edge. B,C) The endothelial cell (EC) has a radius of 10 μm and is 1 μm thick. The smooth muscle cell (SMC) has a radius of 5 μm and is 4 μm thick. The head of the MEJ is spherical with a radius of 0.5 μm that is located on top of a neck that is 1.525 μm long. Stimulation was modeled by increasing the concentration at the green boundaries in the respective cell. Scale bars indicate the respective dimensions. D,E) The model was extended to include endoplasmic reticulum (ER) in the EC. The ER had a radius of 6 μm, 0.5 μm thick and located 0.1 μm from the plasma membrane. The SMC is shown in yellow, the EC in green, the MEJ in red and the ER in orange.

doi:10.1371/journal.pone.0033632.g001

The model consisted of three main regions, the SMC, the EC and the MEJ that protrudes from the EC in the present model. The protrusion is in continuity with the EC and the SMC is connected with the MEJ through gap junctions, which in the model initially were assumed to have a lumped pore radius of 100 nm (Fig. 1B, marked with blue). Chemical species can diffuse between the cells through the gap junctions. In the model the head of the MEJ is spherical with a radius of 0.5 μm and located on top of a neck which is 1.525 μm long and has a radius of 0.125 μm. As shown in Fig. 1 the EC and SMC have different morphology. The EC has a radius of 10 μm and is 1 μm high whereas the SMC has a radius of 5 μm and is 4 μm high. Hence, we have for simplicity assumed that EC and SMC volumes are equal as also suggested in previous work on myoendothelial signaling [16].

Diffusion and chemical reactions were modeled using a diffusion-reaction equation,

$$\frac{\partial c}{\partial t} = \nabla(D_c \nabla c) - R \quad \textbf{(1)}$$

where $c$ is the concentration of the diffusible species, $D_c$ is a diagonal matrix where the elements in the diagonal are the diffusion coefficients for $c$ in the $x$ and $z$ directions, and $R$ expresses chemical reactions and transport e.g. buffer reactions.

The model was solved numerically using the finite element method. The model was implemented in Comsol Multiphysics 4.1 (Comsol AB) [17] and was meshed with triangles using the built-in mesh function. Maximum element size was $5\times10^{-8}$ m, minimum element size $1\times10^{-9}$ m, maximum element growth rate 1.1 and resolution of curvatures 0.2. All the boundaries in the protrusion including the gap junction area had resolution maximum of $5\times10^{-9}$m, minimum $5\times10^{-10}$ m and a maximum growth rate of 1.2. When the radius of the gap junction in Model 2 was reduced the maximum element size was $1\times10^{-9}$ m around the boundary defined by the gap junction. The maximum growth rate defines how much the element size can grow from a region with smaller elements to a region with larger elements. A maximum growth rate of 1.2 means that the element size can increase by 20% from one element to the next.

## Model 1

In Model 1 we simulated diffusion of $Ca^{2+}$ ions in a non-buffered cytosol in order to quantify the basic properties of the structure. Initially the concentration of $Ca^{2+}$ was 0.1 μM in all compartments. We simulated an increase in bulk cytoplasmic $Ca^{2+}$ concentration in either the EC or SMC by increasing the boundary concentrations (Fig. 1C, green lines) by 0.5 μM to a final concentration of 0.6 μM. Unless explicitly stated the diffusion of $Ca^{2+}$ was assumed to be isotropic with a diffusion coefficient of 233 $\mu m^2/s$ [18].

All parameters are listed in Table 2 and initial conditions in Table 3.

Table 2. List of parameters in the models.

doi:10.1371/journal.pone.0033632.t002

| Parameter | Value | Unit | Description | See Reference |
|---|---|---|---|---|
| $k_{on}$ | $50\times10^{6}$ | 1/((mol/l) s) | Rate constant, $Ca^{2+}$ buffer cytosolic | [32] |
| $k_{off}$ | 25 | 1/s | Rate constant, $Ca^{2+}$ buffer cytosolic | [32] |
| $k_{on,ER}$ | $10\times10^{4}$ | 1/((mol/l) s) | Rate constant, $Ca^{2+}$ buffer ER | [21] |
| $k_{off,ER}$ | 200 | 1/s | Rate constant, $Ca^{2+}$ buffer ER | [21] |
| $D_{Ca}$ | 233 | $\mu m^2/s$ | Diffusion coefficient of $Ca^{2+}$ | [18] |
| $D_{buffer}$ | 13 | $\mu m^2/s$ | Diffusion coefficient of $Ca^{2+}$ buffer | [18] |
| $D_{bound}$ | 13 | $\mu m^2/s$ | Diffusion coefficient of $Ca^{2+}$ buffer bound to $Ca^{2+}$ | [18] |
| $D_{IP3}$ | 280 | $\mu m^2/s$ | Diffusion coefficient of $IP_3$ | [18] |
| $a_4$ | $0.2\times10^{6}$ | 1/((mol/l) s) | Rate constant for $IP_3R$ | [22] |
| $a_5$ | $20\times10^{6}$ | 1/((mol/l) s) | Rate constant for $IP_3R$ | [22] |
| $d_1$ | $0.13\times10^{-6}$ | mol/l | Equilibrium constant for $IP_3R$ | [22] |
| $d_2$ | $1.049\times10^{-6}$ | mol/l | Equilibrium constant for $IP_3R$ | [22] |
| $d_3$ | $943.4\times10^{-9}$ | mol/l | Equilibrium constant for $IP_3R$ | [22] |
| $d_4$ | $144.5\times10^{-9}$ | mol/l | Equilibrium constant for $IP_3R$ | [22] |
| $d_5$ | $82.34\times10^{-9}$ | mol/l | Equilibrium constant for $IP_3R$ | [22] |
| $V_{SERCA\_f}$ and $V_{SERCA\_r}$ | $1.022\times10^{-7}$ | mol/(m² s) | SERCA pump, max capacity | [21] |
| $K_{SERCA\_f}$ | $260\times10^{-9}$ | mol/l | SERCA pump, affinity | [21] |
| $H_{SERCA\_f}$ and $H_{SERCA\_r}$ | 0.75 | - | SERCA pump, cooperativity | [21] |
| $K_{SERCA\_r}$ | $1.8\times10^{-3}$ | mol/l | SERCA pump, affinity | [21] |

doi:10.1371/journal.pone.0033632.t002

Table 3. List of variables and initial conditions.

doi:10.1371/journal.pone.0033632.t003

| Variable | Initial value | Unit | Description |
|---|---|---|---|
| $Ca^{2+}$ (cytosol) | 0.1 | μM | $Ca^{2+}$ concentration in the cytosol |
| $Ca^{2+}$ (ER) | 500 | μM | $Ca^{2+}$ concentration in ER |
| $IP_3$ | 0.1 | μM | $IP_3$ concentration |
| Buffer (cytosol) | 83.3 | μM | $Ca^{2+}$ buffer (free form) in the cytosol |
| Buffer (ER) | 57.6 | mM | $Ca^{2+}$ buffer (free form) in ER |
| $Ca^{2+}$, Bound (cytosol) | 16.7 | μM | $Ca^{2+}$ buffer (bound to $Ca^{2+}$) in the cytosol |
| $Ca^{2+}$,Bound (ER) | 14.4 | mM | $Ca^{2+}$ buffer (bound to $Ca^{2+}$) in ER |
| $x_{000}$ | 0.265 | 1 | Fraction of $IP_3R$ on free form |
| $x_{001}$ | 0.185 | 1 | Fraction of $IP_3R$ with $Ca^{2+}$ bound to the activating site |
| $x_{010}$ | 0.324 | 1 | Fraction of $IP_3R$ with $Ca^{2+}$ bound to the inhibitory site |

doi:10.1371/journal.pone.0033632.t003

## Model 2

The model was extended to include the effect of $Ca^{2+}$-induced $Ca^{2+}$-release (CICR) from *endoplasmic reticulum* (ER) in the EC and MEJ. This model contains the following diffusible species: $Ca^{2+}$ in the cytosol and in ER, $Ca^{2+}$ buffers in the cytosol and ER and $IP_3$ in the cytosol (see Table 2 for diffusion coefficients). We assumed that all pumps and channels were distributed uniformly in the membranes of the ER and hence any effects from point sources were neglected. The structure of the ER inside the MEJ was based on electron microscopic images [7]. The ER in the EC had a radius of 6 μm and was 0.5 μm thick and 0.1 μm from the upper part of the EC membrane. In the inclined part of the MEJ neck the latter distance was 60 nm and in the vertical part it was 75 nm. In the head of the MEJ the ER had a radius of 0.25 μm (Fig. 1D–E).

We also included the reversible binding of $Ca^{2+}$ to $Ca^{2+}$ buffers in the cytosol of the EC, MEJ and SMC,

$$Ca^{2+}_{cyt} + Buffer_{cyt} \underset{koff}{\overset{kon}{\rightleftarrows}} Ca^{2+},bound_{cyt} \quad (2)$$

with the two rate constants $k_{on}$ and $k_{off}$ (see Table 2). The reaction was modeled using first and second order kinetics.

$$R_{buffer,cyt} = k_{off} \cdot \left[Ca^{2+},bound_{cyt}\right] - k_{on} \cdot \left[Ca^{2+}_{cyt}\right] \cdot \left[Buffer_{cyt}\right] \quad (3)$$

The ratio between $k_{off}$ and $k_{on}$ gives the dissociation constant Kd (Kd $= k_{off}/k_{on}$), which corresponds numerically to the $Ca^{2+}$ concentration at which the buffer is 50% saturated. In the cytosol of the SMC, EC and MEJ the Kd was 0.5 μM [19], [20]. The total buffer concentration was 0.1 mM. $Ca^{2+}$ buffer binding was also modeled in ER:

$$Ca^{2+}_{ER} + Buffer_{ER} \underset{koff,ER}{\overset{kon,ER}{\rightleftarrows}} Ca^{2+},bound_{ER} \quad (4)$$

which was modeled as in the cytosol, see Eq. $R_{buffer,ER} = k_{off,ER} \cdot \left[Ca^{2+},bound_{ER}\right] - k_{on,ER} \cdot \left[Ca^{2+}_{ER}\right] \cdot \left[Buffer_{ER}\right]$ (5) where the total Ca2+ buffer capacity was 72 mM, with a Kd value of 2 mM [21]. We modeled CICR using the model by De Young and Keizer [22] which we adapted to the present model. The flux of Ca2+ through the IP3 receptor channel (IP3R) was dependent on the permeability of the channel (PIP3R), the fraction of channels being in an open state $(x^3_{110})$ and the Ca2+ gradient across the ER membrane. The flux across the ER membrane was then:

$$J_{IP3R} = P_{IP3R} \cdot x^3_{110} \left( \left[Ca^{2+}\right]_{ER} - \left[Ca^{2+}\right]_{cyt} \right). \quad (6)$$

The $IP_3$ receptor is activated by $IP_3$ and by low concentrations of $Ca^{2+}$, whereas $Ca^{2+}$ at higher concentrations inhibits the channel. Using a rapid equilibrium approximation De Young and Keizer [22] presented a set of equations to describe the different states of the receptor. Each position in the subscript ($x_{ijk}$) describes if a ligand is bound (1) or not bound (0) to the three independent binding sites. The first place (i) refers to binding of $IP_3$, the second (j) refers to binding of $Ca^{2+}$ to the activating site and the third (k) to binding of $Ca^{2+}$ to the inhibitory site.

In the model it is assumed that the active channel is composed of three subunits and because it is assumed that all three have to be in the state $x_{110}$ for the channel to be open, the fraction of open channels is given by $x_{110}^3$. The dynamical change between the forms without $IP_3$ is described with the following equations [22];

$$\frac{dx_{000}}{dt} = -V_1 - V_3 \quad (7)$$

$$\frac{dx_{001}}{dt} = V_1 - V_4 \quad (8)$$

$$\frac{dx_{010}}{dt} = V_3 - V_2 \quad (9)$$

$$x_{011} = 1 - x_{000} - x_{001} - x_{010} \quad (10)$$

where $V_1 = a_4([Ca^{2+}]_{cyt} x_{000} - d_4 x_{001})$ (11)

$$V_2 = a_4([Ca^{2+}]_{cyt} x_{010} - d_4 x_{011}) \quad (12)$$

$$V_3 = a_5([Ca^{2+}]_{cyt} x_{000} - d_5 x_{010}) \quad (13)$$

$$V_4 = a_5([Ca^{2+}]_{cyt} x_{001} - d_5 x_{011}) \quad (14)$$

and the $IP_3$ bound forms are found as: $x_{100} = ([IP_3]/d_1) x_{000}$ (15)

$$x_{101} = ([IP_3]/d_3) x_{001} \quad (16)$$

$$x_{110} = ([IP_3]/d_1) x_{010} \quad (17)$$

$$x_{111} = ([IP_3]/d_3) x_{011} \quad (18)$$

where $a_x$ and $d_x$ are constants (Table 2).

Re-uptake of $Ca^{2+}$ into the ER by the SERCA pump (a $Ca^{2+}$ pump) was modeled using [21].

$$J_{SERCA} = \frac{V_{SERCA\_f}\left([Ca^{2+}]_{cyt}/k_{SERCA\_f}\right)^{H_{SERCA\_f}} - V_{SERCA\_r}\left([Ca^{2+}]_{cyt}/k_{SERCA\_r}\right)^{H_{SERCA\_r}}}{1+\left([Ca^{2+}]_{cyt}/k_{SERCA\_f}\right)^{H_{SERCA\_f}} + \left([Ca^{2+}]_{ER}/k_{SERCA\_r}\right)^{H_{SERCA\_r}}} \quad \textbf{(19)}$$

where the $f$ and $r$ indicates the forward and reverse modes. All constants are given in Table 2and initial conditions in Table 3.

The $IP_3$ concentration was 0.1 μM prior to stimulation and the maximum $Ca^{2+}$ permeability through the $IP_3R$ channel ($P_{IP3R}$) was fitted to counterbalance the activity of the SERCA pump[21]. To simulate an increase in $IP_3$ in the EC or the SMC we increased the concentration of $IP_3$ to 0.4 μM around all external boundaries in either the EC or the SMC (Fig. 1C). Note that the diffusion coefficient of $IP_3$ is higher than that of $Ca^{2+}$ (Table 2).

## Results

To elucidate the effect of the MEJ on diffusion between the two cell types, we considered a simple system of $Ca^{2+}$ diffusion in an unbuffered system and we evaluated two scenarios where either the EC or the SMC was stimulated by increasing the bulk cytosolic $Ca^{2+}$ concentration (Model 1). An increase in bulk $Ca^{2+}$ concentration was simulated by raising the concentration of cytosolic $Ca^{2+}$ ($[Ca^{2+}]_{cyt}$) by 0.5 μM above the resting level at all boundaries in either of the cells, except for the boundaries of the MEJ (see Fig. 1 for details), and we assumed initially that the cytosol was unbuffered. Following perturbation of the $Ca^{2+}$ concentration, the bulk cytosolic concentration increased to almost 0.6 μM within 10 ms in the EC and within 30 ms in the SMC (Fig. 2A and B). The slower increase in the SMC is due to the lower surface to volume ratio. The fast increase in concentration shows that raising the concentration at the boundaries was a good approximation of simulating a global perturbation in the cell.

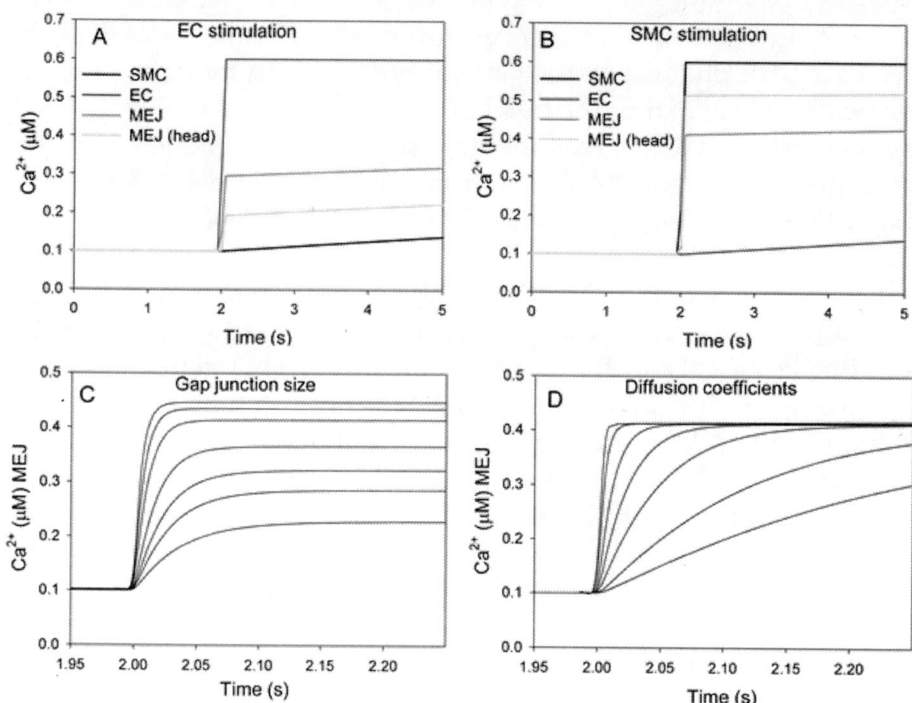


**Figure 2. Model 1, diffusion through the MEJ.**

The effect of the MEJ on diffusion between the SMC and EC was tested by increasing the $Ca^{2+}$ concentration in A): EC from 0.1 μM to 0.6 μM and B): in the SMC from 0.1 μM to 0.6 μM at time 2 s. The average concentration in the EC, the SMC, the MEJ and only in the head of the MEJ is shown in both cases. C) Average $Ca^{2+}$ concentrations in the MEJ as a function of gap junction pore size. From below pore size was: 10, 20, 30, 50, 100, 150 and 200 nm. D) The diffusion coefficient was changed within a chemical/biological realistic regime and the average $Ca^{2+}$ concentration in the MEJ was found for each value. From below 10, 20, 50, 100, 200, 500 and 1000 $\mu m^2/s$.

doi:10.1371/journal.pone.0033632.g002

While the $Ca^{2+}$ response in the stimulated cell was very fast, the bulk cytosolic $Ca^{2+}$concentration in the other cell responded very slowly with only small changes in bulk cytosolic $Ca^{2+}$ levels even after several seconds (Fig. 2A and B). Whereas the responses in the bulk cytosolic $Ca^{2+}$ concentrations were quite similar in the two cases, the changes in the $Ca^{2+}$concentration in the microdomain formed by the MEJ and the head of the MEJ differed substantially. It is noteworthy that whereas the $Ca^{2+}$ signal that originated from the EC only had a modest effect on the head and the neck of the MEJ (Fig. 2A), the $Ca^{2+}$ signal from the SMC increased the $Ca^{2+}$ level in the head of MEJ, which is a part of the EC, to >0.5 μM within 40 ms (Fig. 2B). This shows that a signal from the cell that makes contact with the MEJ, here the SMC, can be transmitted very rapidly to the other cell, but less easily in the opposite direction. The rectification was caused by the direct coupling of the MEJ head to the SMC. Because of the geometry $Ca^{2+}$ diffuses easily from the SMC to the MEJ head. Since the bulk of the EC is separated from the MEJ by a narrow neck the efflux from the MEJ head into the bulk of the EC is delayed. The SMC can therefore easily drive an increase in MEJ $Ca^{2+}$concentration. In contrast, a concentration increase in the bulk cytoplasm of the EC will only lead to a slow increase in the concentration in the head of the MEJ. In addition compounds that diffuse from the EC into the MEJ head will easily diffuse into the SMC, and the effect on the concentration in the head will be minimized.

The permeability of the gap junction and the diffusion coefficient could be central features of the model. We therefore tested how the increase in MEJ $Ca^{2+}$ concentration in response to an increase in the SMC $Ca^{2+}$ concentration, varied with changes in gap junction size (i.e. the degree of coupling) (Fig. 2C) and diffusion coefficient (Fig. 2D). We found that the steady-state concentration in the MEJ decreased when the gap junction size was lowered, whereas the rise time was almost unaffected. In contrast, a general decrease in the diffusion coefficient prolonged the rise time for the increase in MEJ concentration, but the steady state concentration was unaffected (Fig. 2D).

We next considered the presence of an amplifier in the MEJ by including a model of the *endoplasmic reticulum* (ER) (Model 2). ER is present in the MEJ and it may extend into the cytosol of the EC [6], [7]. Although it is unknown how often this is the case, we assumed in the model that the ER in the MEJ couples to the ER in the cytosol (Fig. 1D and E). The ER surface, including that inside the MEJ, contains $IP_3$ receptor channels ($IP_3R$) [6], [7] enabling $Ca^{2+}$ induced

$Ca^{2+}$ release (CICR) [23]. When $IP_3$ was increased in the SMC there was a pronounced $Ca^{2+}$ release in the MEJ after less than 100 ms due to $Ca^{2+}$ release from the ER (Fig. 3). MEJ $Ca^{2+}$ concentration increased rapidly (<50 ms) after SMC $IP_3$ was an increased (Fig. 3A) and the average $Ca^{2+}$ concentration in the MEJ remained high for more than 1 s (Fig. 3B). When an $IP_3$ increase was generated in the SMC, $Ca^{2+}$ from the ER in the EC will elevate the EC $[Ca^{2+}]_{cyt}$, (Fig. 4A), but the reverse flux of $Ca^{2+}$ from the EC to the SMC will only have a marginal effect on SMC $[Ca^{2+}]_{cyt}$ (Fig. 4A).

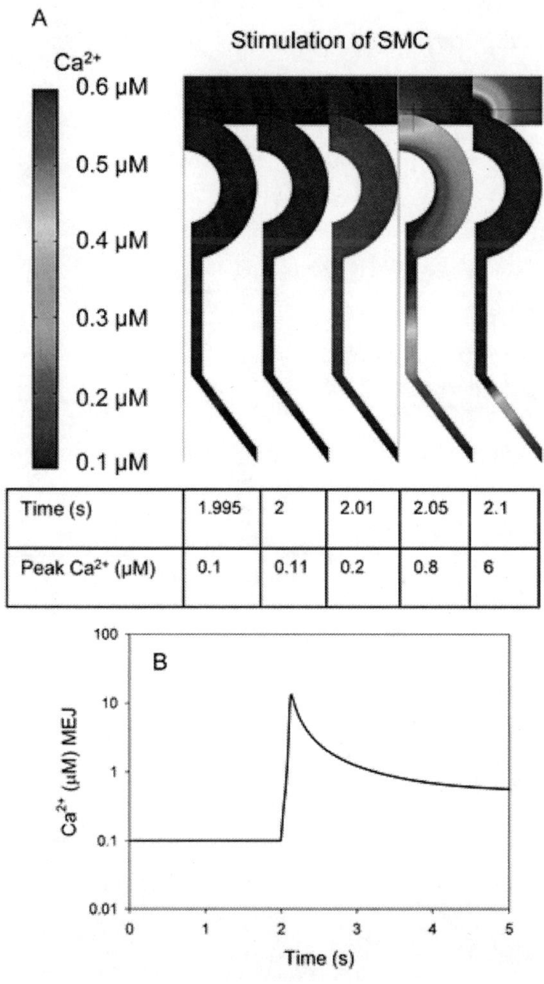


Figure 3. Model 2, the MEJ is a sensor of the adjacent cell.

A) The changes in $Ca^{2+}$ concentration in the protrusions following an increase in the boundary level of $IP_3$ in the SMC from 0.1 to 0.4 μM. The local $Ca^{2+}$ cloud in the SMC is due to diffusion of $Ca^{2+}$ from the EC/MEJ through the gap junction. The concentration of $Ca^{2+}$ in the protrusion is shown at the indicated time points. Below the bar with the time points are the peak concentrations of $Ca^{2+}$. The $IP_3$ level was increased at time 2 s in the SMC. B) Average $Ca^{2+}$ concentration in the MEJ following an $IP_3$ increase in the SMC. When $IP_3$ was increased from 0.1 to 0.4 μM in the SMC the $Ca^{2+}$ microdomain in the MEJ was on average of 4 μM at time 2.1 s and 12 μM at time 2.14 s.

doi:10.1371/journal.pone.0033632.g003

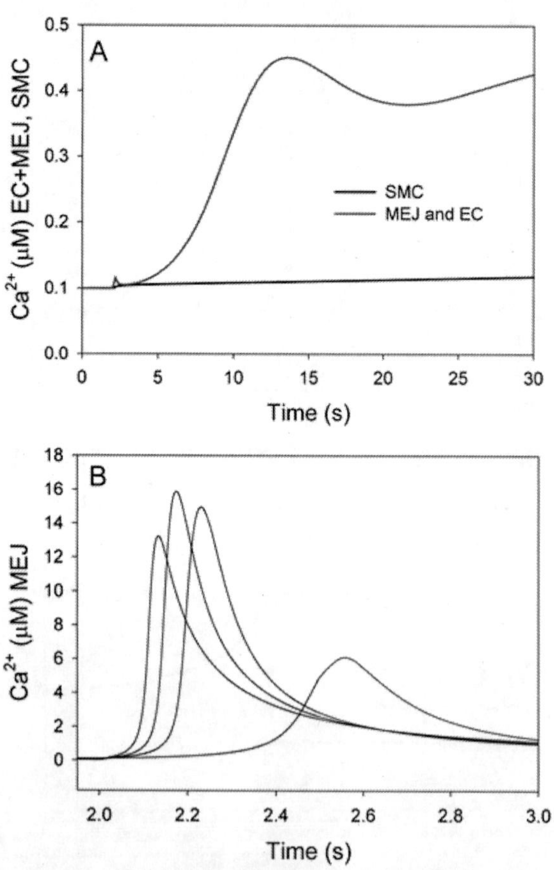


Figure 4. The MEJ amplifies signals from the SMC to the EC.

Following an increase in SMC $IP_3$ concentration from 0.1 to 0.4 μM, $Ca^{2+}$ increased in the MEJ. A) When $IP_3$ was increased to 0.4 μM in the SMC the global bulk EC $[Ca^{2+}]_{cyt}$ increased to >0.3 μM within 5–10 s in the EC, but the $[Ca^{2+}]_{cyt}$ in the SMC remained unaffected. B) The radius of the pore was set to the following values (curves from the right): 10, 30, 50 and 100 nm, and average MEJ concentration was calculated following an increase in $IP_3$ in the SMC.

doi:10.1371/journal.pone.0033632.g004

The permeability of the gap junctions in the MEJ is to the best of our knowledge not known, and to address the impact of that we tested the model for various sizes of the gap junction pore (Fig. 2C and Fig. 4B). We found that over a range of gap junction sizes from 10–200 nm an increase in $IP_3$ in the SMC was able to generate a substantial $Ca^{2+}$ signal in the MEJ as seen by the increase in MEJ $Ca^{2+}$ concentration (Fig. 4B). The simulation when the pore was 200 nm was similar to the simulation where the pore was 100 nm (not shown).

We addressed the possibility of variation in the length of the MEJ by modeling changes in the length of the longitudinal part of the neck that connects the head with the EC body. For this purpose we used Model 1 that only includes one compound ($Ca^{2+}$). To model variations in the length of the neck the diffusion coefficient was re-scaled in the direction of the z-axis in the narrow part of the neck (Fig. 5). Changes in length will scale with the diffusion coefficient multiplied by the inverse of the scaling factor raised to the power of two, i.e. quadrupling the diffusion coefficient corresponds to halving the length. To quantify the effect of changing the length of the neck we simulated the average $Ca^{2+}$ concentration in the head of the MEJ 0.1 s after stimulation of one of the cell types. When the EC was stimulated a decreased length of the neck increased the $Ca^{2+}$ concentration in the head of the MEJ after stimulation whereas a longer neck decreased the concentration (Fig. 5A). When the SMC was stimulated a longer neck increased the $Ca^{2+}$ concentration in the head of the MEJ (Fig. 5B). Hence, a longer neck will increase the polarization effect. If the MEJ was removed from the model and the EC coupled directly to the SMC, we only observed a minor change of 0.018 μM after 0.1 s in the cell opposite of the one that was activated (not shown). Hence, without the local environment provided by the MEJ a signal from the SMC is diluted in the EC. We have also tested the effect of re-scaling the neck in the model including CICR (Model 2, not shown) and we found that increasing the length of the neck with a factor of 2–3 still allowed an $IP_3$ signal from the SMC to initiate a global $Ca^{2+}$ signal in the EC within a few seconds. The models predict that for different lengths of the neck the structure can rectify a signal between the two cells and increasing the length of the neck will increase that effect.

## DISCUSSION

ECs and SMCs are often connected by gap junctions that are located on MEJs. The distribution of ion channels near the MEJ appears to be tightly regulated and there is evidence that the local environment i.e. ion channel localization is highly controlled [24]. We have investigated the impact of the structure of the MEJ using a 2D axi-symmetric mathematical model and we found that the structure of the MEJ itself can influence the signaling between the two cell types by rectifying information flow.

We perturbed the model by simulating an increase in the bulk cytosolic concentration of a diffusible species in either the EC or the SMC. The increase in concentration was implemented as a concentration increase at the boundaries in the respective cells (Fig. 1C). The lag time between bulk cytosolic concentration had reached the level at the boundary from the time it was changed was only 10–30 ms. Since we expect signals originating from the lumen of the vessel to primarily act on the part of the EC membrane facing the lumen, and to a much lesser extent affect the cell membrane in the sequestered region of the MEJ, we deliberately excluded the MEJ from the rise in the cytosolic $Ca^{2+}$ concentration. The main conclusion from the results shown in Figure 2 is that for a small diffusible species (in this case $Ca^{2+}$) that can pass through gap junctions, the structure of the MEJ leads to significant signaling rectification in the sense that the concentration of the diffusible species in the head responds more rapidly to changes in the SMC than it does to changes in the EC from which it originates.

The degree of asymmetry depends on two factors, the degree of gap junctional coupling between the head and the SMC and the length of the neck region that connects the head to the bulk of the EC. As can be seen in Figure 2C, increasing the degree of coupling increases the steady state concentration achieved in the head, but only has a minor effect on the time it takes to reach this level. This is to be expected, since in the model, the steady state concentration will be determined by a balance between the influx from the SMC and the loss of through the neck region. The importance of the length of the neck region is illustrated in Figures 5A and B where we simulated various lengths of the neck region by varying the magnitude of the diffusion coefficient in the direction of the long axis of the neck region. Imposing anisotropic diffusion conditions is equivalent to changing the length of the neck region. A longer neck region will impede the diffusion of solutes from the bulk cytosol of the EC into the head of the MEJ. This effect can be seen in Figure 5A which shows the $Ca^{2+}$ concentration in the head 0.1 s after stimulation of the EC. It is clear that the longer the neck, the smaller the concentration obtained. The reverse was true for the response of the head to an increase in SMC $Ca^{2+}$concentration, where a longer neck hindered the loss of $Ca^{2+}$ from the head, and therefore led to a higher concentration following the stimulation (Fig. 5B). The results clearly indicate that the micro-anatomy of biological systems may have an important role in signal transduction and modulation.

We next addressed the question of how a change in head concentration of a given substance may affect the EC. One possibility could be the presence of an amplification mechanism within the head region. It is well known that the ER may extend into the MEJ, and we therefore considered the possible role of the ER in signal amplification. In this scenario we modeled the diffusion of $IP_3$, and as in the former case, the head region quickly equilibrated with respect to changes in $IP_3$ levels in the SMC. Because of the presence of the ER, the rise in the $IP_3$ level in the head will lead to $Ca^{2+}$ release from the ER through the $IP_3$ receptors. Such a localized $Ca^{2+}$ release may then spread into the EC as a $Ca^{2+}$ wave mediated by $Ca^{2+}$ induced $Ca^{2+}$release (Fig. 3). We hypothesize that such an amplification mechanism may effectively propagate signals from the SMC into the EC. Indeed, recent experiments have suggested that $IP_3$ from the SMC can diffuse through the gap junctions in the MEJs and induce a local $Ca^{2+}$increase in the EC [25], [26].

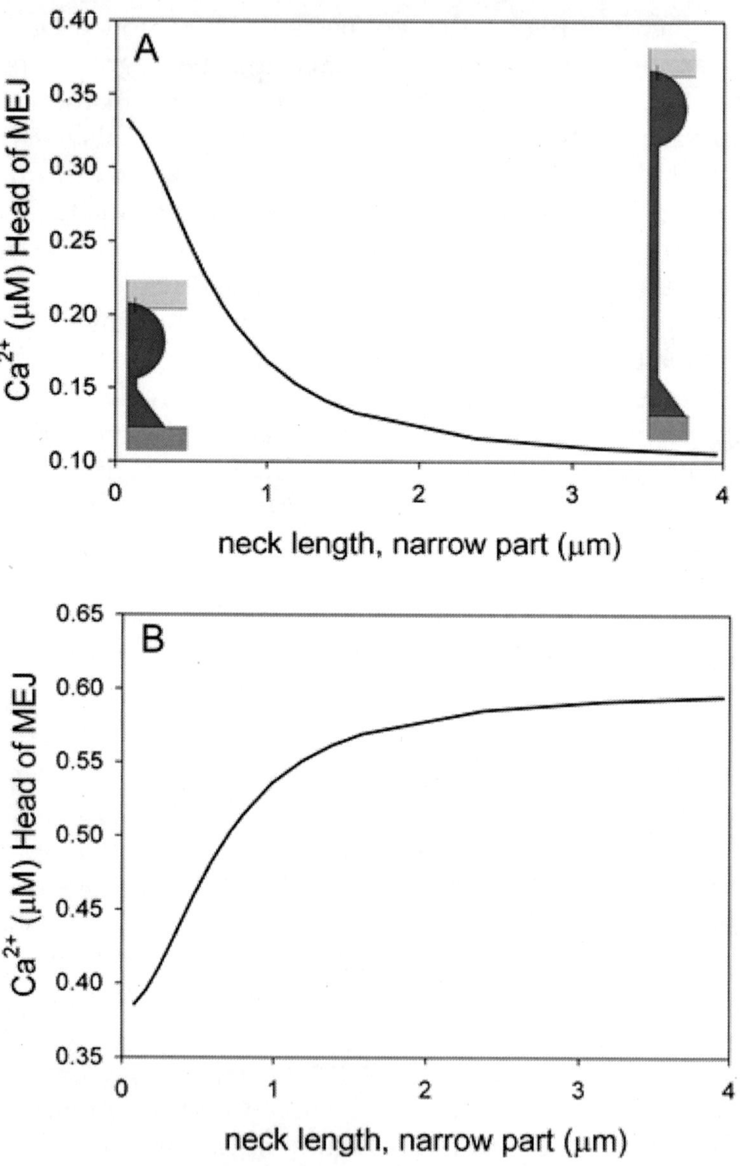


Figure 5. The length of the MEJ.

The effect of changing the length of the narrow/vertical part of the neck of the myoendothelial junction was tested on the model only including $Ca^{2+}$ (Model 1). Re-scaling the length of the neck was implemented by re-scaling the diffusion coefficient in the z-axis in the neck. The value used in Fig. 1, 2, 3, and 4 was 1.26 μm. To quantify the effect of neck re-scaling, the average bulk $Ca^{2+}$ concentration in the head of the myoendothelial junction was measured 0.1 s after stimulation when A): the EC was stimulated and B): when the SMC was stimulated.

doi:10.1371/journal.pone.0033632.g005

The restricted space in the MEJ is an example of a local $Ca^{2+}$ microdomain, i.e. a restricted volume of the cytoplasm where the local $Ca^{2+}$ concentration may reach high levels (Fig. 2 and3). Another example of a $Ca^{2+}$ microdomain is the restricted space between endoplasmic/sarcoplasmic reticulum and the plasma membrane [27]–[29], close to ion channels [30], [31]. The wrinkled structures of the plasma membrane in neutrophil granulocytes have also been suggested to give rise to $Ca^{2+}$ microdomains where the local concentration of $Ca^{2+}$ can reach values of more than 10 μM [32], [33].

$Ca^{2+}$ microdomains have important roles in regulation of processes near membranes including regulation of ion channels [27]–[31]. In small arteries $Ca^{2+}$ microdomains in the SMCs regulate the activity of BK channels and thereby play a central role in regulation of the tone [29], [34]. In the ECs from small arteries the spatial distribution of $K^+$ channels is tightly regulated [24]. The $K^+$ channel type KCa3.1 is highly expressed in the MEJ [8], [24], [35] and KCa2.3 is present in the MEJ and at the cell border [24]. KCa3.1 and KCa2.3 are both activated by $Ca^{2+}$ in the range 50–900 nM [36], and hence an increase in MEJ $Ca^{2+}$ concentration could activate $K^+$currents in the MEJ, which in turn would hyperpolarize the entire cell. Activation of the channel would in addition lead to an increase in the $K^+$ concentration in the internal elastic lamina, which could, in principle, activate Kir channels in both EC and SMCs. It is known that an increase in interstitial $K^+$ concentration dilates vessels in part due to Kir activation [37]–[40]. It is therefore plausible that activation of $K^+$ channels in the MEJ could lead to a hyperpolarization of the SMC, in turn inhibiting voltage sensitive $Ca^{2+}$ channels leading to vasodilation.

It has previously been suggested that asymmetrical distribution of $IP_3Rs$ and $IP_3$ phosphatases also can give rise to an asymmetrical $IP_3/Ca^{2+}$ signaling between ECs and SMCs [14], [41], and we speculate that these effects could act together with the structural effects presented here. Knowledge of the distribution of $IP_3Rs$ in and near MEJ in arterioles from various tissues is limited. However, our conclusions are not dependent on any functional polarization, but only depend on the actual structure of the MEJ. If degradation of $IP_3$ by phosphatases is taken into account it will enhance the reported effect of a unidirectional flow from the SMC to the EC, because $IP_3$ that enters the MEJ from the SMC has a relative short distance to travel before it encounters the ER in the MEJ. $IP_3$ produced in the EC will on the other hand have to travel through the entire MEJ before entering the SMC and this would decrease the amount of $IP_3$entering the SMC. In the SMC $IP_3$ could activate opening of $IP_3$Rs in the sarcoplasmic reticulum, but as we have shown in Figures 2 and 3 the flow of molecules from the EC to the SMC is operating on a time scale that is substantially slower than the flow from the SMC to the EC. Hence, under normal conditions it would be unlikely that $IP_3$ produced in the EC could activate $IP_3$R channels in the SMC. If phosphatases in both ECs and SMCs were inhibited and the EC was stimulated for a prolonged period it could be speculated that $IP_3$ from the EC could stimulate a $Ca^{2+}$ release in the SMC.

In conclusion, the micro-anatomical structure of the MEJ, with a long neck region restricting diffusion between the head and the bulk of the cytosol, by itself leads to a rectification of information flow between the SMC and the EC. Changes in $IP_3$ and $Ca^{2+}$ in the SMC are rapidly and efficiently transmitted to the EC, whereas the reverse is not the case.

## Author Contributions

Conceived and designed the experiments: JCB. Performed the experiments: JCB. Analyzed the data: JCB JCBJ NHHR. Contributed reagents/materials/analysis tools: JCB. Wrote the paper: JCB JCBJ NHHR.

## REFERENCES

1. Kotaleski JH, Blackwell KT (2010) Modelling the molecular mechanisms of synaptic plasticity using systems biology approaches. Nat Rev Neurosci 11: 239–251.

2. Smith PD, Brett SE, Luykenaar KD, Sandow SL, Marrelli SP, et al. (2008) KIR channels function as electrical amplifiers in rat vascular

smooth muscle. J Physiol 586: 1147–1160.

3. Chadha PS, Liu L, Rikard-Bell M, Senadheera S, Howitt L, et al. (2011) Endothelium-dependent vasodilation in human mesenteric artery is primarily mediated by myoendothelial gap junctions intermediate conductance calcium-activated K+ channel and nitric oxide. J Pharmacol Exp Ther 336: 701–708.
4. Sandow SL, Senadheera S, Bertrand PP, Murphy TV, Tare M (2011) Myoendothelial contacts, gap junctions and microdomains: anatomical links to function? Microcirculation In press.
5. Sandow SL, Hill CE (2000) Incidence of myoendothelial gap junctions in the proximal and distal mesenteric arteries of the rat is suggestive of a role in endothelium-derived hyperpolarizing factor-mediated responses. Circ Res 86: 341–346.
6. Ledoux J, Taylor MS, Bonev AD, Hannah RM, Solodushko V, et al. (2008) Functional architecture of inositol 1,4,5-trisphosphate signaling in restricted spaces of myoendothelial projections. Proc Natl Acad Sci U S A 105: 9627–9632.
7. Sandow SL, Haddock RE, Hill CE, Chadha PS, Kerr PM, et al. (2009) What's where and why at a vascular myoendothelial microdomain signalling complex. Clin Exp Pharmacol Physiol 36: 67–76.
8. Mather S, Dora KA, Sandow SL, Winter P, Garland CJ (2005) Rapid endothelial cell-selective loading of connexin 40 antibody blocks endothelium-derived hyperpolarizing factor dilation in rat small mesenteric arteries. Circ Res 97: 399–407.
9. Unger VM, Kumar NM, Gilula NB, Yeager M (1999) Three-dimensional structure of a recombinant gap junction membrane channel. Science 283: 1176–1180.
10. Evans WH, Martin PE (2002) Gap junctions: structure and function (Review). Mol Membr Biol 19: 121–136.
11. Isakson BE, Ramos SI, Duling BR (2007) Ca2+ and inositol 1,4,5-trisphosphate-mediated signaling across the myoendothelial junction. Circ Res 100: 246–254.
12. Dora KA, Doyle MP, Duling BR (1997) Elevation of intracellular calcium in smooth muscle causes endothelial cell generation of NO in arterioles. Proc Natl Acad Sci U S A 94: 6529–6534.
13. de Wit C, Hoepfl B, Wolfle SE (2006) Endothelial mediators and communication through vascular gap junctions. Biol Chem 387: 3–9.
14. Isakson BE (2008) Localized expression of an Ins(1,4,5)P-3 receptor at the myoendothelial junction selectively regulates heterocellular Ca2+ communication. Journal of Cell Science 121: 3664–3673.

15. Dora KA, Sandow SL, Gallagher NT, Takano H, Rummery NM, et al. (2003) Myoendothelial gap junctions may provide the pathway for EDHF in mouse mesenteric artery. J Vasc Res 40: 480–490.
16. Beny JL (1999) Information Networks in the Arterial Wall. News Physiol Sci 14: 68–73.
17. 21–2–2012: Comsol Multiphysics 4.1 (Comsol AB) Available:http://www.comsol.com/products/multiphysics/Accessed.
18. Allbritton NL, Meyer T, Stryer L (1992) Range of messenger action of calcium ion and inositol 1,4,5-trisphosphate. Science 258: 1812–1815.
19. Kapela A, Bezerianos A, Tsoukias NM (2008) A mathematical model of Ca2+ dynamics in rat mesenteric smooth muscle cell: agonist and NO stimulation. J Theor Biol 253: 238–260.
20. Silva HS, Kapela A, Tsoukias NM (2007) A mathematical model of plasma membrane electrophysiology and calcium dynamics in vascular endothelial cells. Am J Physiol Cell Physiol 293: C277–C293.
21. Means S, Smith AJ, Shepherd J, Shadid J, Fowler J, et al. (2006) Reaction diffusion modeling of calcium dynamics with realistic ER geometry. Biophys J 91: 537–557.
22. De Young GW, Keizer J (1992) A single-pool inositol 1,4,5-trisphosphate-receptor-based model for agonist-stimulated oscillations in Ca2+ concentration. Proc Natl Acad Sci U S A 89: 9895–9899.
23. Berridge MJ (1993) Inositol Trisphosphate and Calcium Signaling. Nature 361: 315–325.
24. Dora KA, Gallagher NT, McNeish A, Garland CJ (2008) Modulation of endothelial cell KCa3.1 channels during endothelium-derived hyperpolarizing factor signaling in mesenteric resistance arteries. Circ Res 102: 1247–1255.
25. Nausch LW, Bonev AD, Heppner TJ, Tallini Y, Kotlikoff MI, et al. (2012) Sympathetic nerve stimulation induces local endothelial Ca2+ signals to oppose vasoconstriction of mouse mesenteric arteries. Am J Physiol Heart Circ Physiol 302: H594–H602.
26. Tran CH, Taylor MS, Plane F, Nagaraja S, Tsoukias NM, et al. (2012) ENDOTHELIAL Ca2+ WAVELETS AND THE INDUCTION OF MYOENDOTHELIAL FEEDBACK. Am J Physiol Cell Physiol In press.
27. Lederer WJ, Niggli E, Hadley RW (1990) Sodium-calcium exchange in excitable cells: fuzzy space. Science 248: 283.
28. Berridge MJ (2006) Calcium microdomains: organization and function. Cell Calcium 40: 405–412.
29. Brenner R, Perez GJ, Bonev AD, Eckman DM, Kosek JC, et al.

(2000) Vasoregulation by the beta1 subunit of the calcium-activated potassium channel. Nature 407: 870–876.

30. Naraghi M, Neher E (1997) Linearized buffered Ca2+ diffusion in microdomains and its implications for calculation of [Ca2+] at the mouth of a calcium channel. J Neurosci 17: 6961–6973.
31. Chad JE, Eckert R (1984) Calcium domains associated with individual channels can account for anomalous voltage relations of CA-dependent responses. Biophys J 45: 993–999.
32. Brasen JC, Olsen LF, Hallett MB (2010) Cell surface topology creates high Ca2+ signalling microdomains. Cell Calcium 47: 339–349.
33. Davies EV, Hallett MB (1996) Near membrane Ca2+ changes in neutrophils. Biochem Soc Trans 24: 92S.
34. Jaggar JH, Porter VA, Lederer WJ, Nelson MT (2000) Calcium sparks in smooth muscle. Am J Physiol Cell Physiol 278: C235–C256.
35. Sandow SL, Neylon CB, Chen MX, Garland CJ (2006) Spatial separation of endothelial small- and intermediate-conductance calcium-activated potassium channels (K(Ca)) and connexins: possible relationship to vasodilator function? J Anat 209: 689–698.
36. Hille , Bertil (2001) Ion channels of excitable membranes. Sunderland, Mass: Sinauer.
37. McCarron JG, Halpern W (1990) Potassium dilates rat cerebral arteries by two independent mechanisms. Am J Physiol 259: H902–H908.
38. Chilton L, Smirnov SV, Loutzenhiser K, Wang X, Loutzenhiser R (2011) Segment-specific differences in the inward rectifier K+ current along the renal interlobular artery. Cardiovasc Res 92: 169–177.
39. Burns WR, Cohen KD, Jackson WF (2004) K+-induced dilation of hamster cremasteric arterioles involves both the Na+/K+-ATPase and inward-rectifier K+ channels. Microcirculation 11: 279–293.
40. Haddy FJ, Vanhoutte PM, Feletou M (2006) Role of potassium in regulating blood flow and blood pressure. Am J Physiol Regul Integr Comp Physiol 290: R546–R552.
41. Kapela A, Bezerianos A, Tsoukias NM (2009) A mathematical model of vasoreactivity in rat mesenteric arterioles: I. Myoendothelial communication. Microcirculation 16: 694–713.

# Chapter 8

# AN EFFICIENT 3D CELL CULTURE METHOD ON BIOMIMETIC NANOSTRUCTURED GRIDS

Maria Wolun-Cholewa[1*], Krzysztof Langer[2], Krzysztof Szymanowski[3], Aleksandra Glodek[1], Anna Jankowska[1], Wojciech Warchol[4], Jerzy Langer[2]

[1]Department of Cell Biology, Poznan University of Medical Science, Poznan, Poland, [2]Laboratory for Materials Physicochemistry and Nanotechnology, Adam Mickiewicz University, Srem, Poland

[3]Department of Mother's and Child's Health, Poznan University of Medical Science, Poznan, Poland

[4]Department of Biophysics, Poznan University of Medical Science, Poznan, Poland

## ABSTRACT

Current techniques of *in vitro* cell cultures are able to mimic the *in vivo* environment only to a limited extent, as they enable cells to grow only in two dimensions. Therefore cell culture approaches should rely on scaffolds that provide support comparable to the extracellular matrix. Here we demonstrate the advantages of novel nanostructured three-dimensional grids fabricated using electro-spinning technique, as scaffolds for cultures of neoplastic cells. The results of the study show that the fibers allow for a dynamic growth of HeLa cells, which form multi-layer structures of symmetrical and spherical character. This indicates that the applied scaffolds are nontoxic and allow proper flow of oxygen, nutrients, and growth factors. In addition, grids have been proven to be useful in *in situ* examination of cells ultrastructure.

## INTRODUCTION

Cultures of human or animal cells in *in vitro* conditions are usually performed for their identification, proliferation or cell death assessment. Frequently, initial evaluations are carried out in two-dimensional (2D) cell culture systems. Despite their widespread use, observations made in 2D systems do not correspond to the results of *in vivo* studies [1]. Thus, currently, researchers are using three dimensional (3D) cultures that mimic the *in vivo* environment more accurately [2]. Cells grown in three-dimensional cultures are more valid targets for discovering and testing of new drugs for cancer treatment [3]. Contrary to 2D culture systems, 3D *in vitro*models have the potential to provide insights into cellular functions such as: differentiation, migration, and gene expression in a controlled and well-defined manner [4]–[6]. 3D cell culture main feature is the ability to mimic the extracellular matrix (ECM) conditions. Materials used for 3D cell culture systems production include both natural and synthetic biopolymers. Lately also biodegradable materials were introduced in scaffold construction, however it was proven that their stability in liquid environment is limited [5], [7]–[14].

One of the methods for fiber-based 3D scaffolds production is electro-spinning [9]. An advantage of those nanostructured grids, which differentiates them from other types of scaffolds, is the reduced diameter of pore size [15]–[16]. Moreover, electrospun scaffolds are built from very small fibers which create large surface areas [1], [9], [13], [15]. This allows for more accurate evaluation of cell proliferation [9], [11], [17].

In this study, electrospun nanostructured fibers in the form of spatial nanostructured 3D grids, fabricated from a polymer mixture, including polyaniline, were examined as a potential tool for 3D culture *in vitro*.

# MATERIALS AND METHODS

## Synthesis of Polyaniline

6.9 g of aniline hydrochloride was dissolved in 300 ml of 1 M HCl. The solution of aniline hydrochloride was combined with a solution of 11.4 g ammonium persulphate dissolved in 200 ml of 1 M HCl. The mixture was left at room temperature for 12 hours with continuous mixing. Next the product was filtered, washed with water, methanol and chloroform. Polyaniline (PANI) was transferred into 200 ml of chloroform and dispersed to a suspension using ultrasound bath.

## Fabrication of Nanofibers using Electro-spinning

150 mg of polystyrene was dissolved in 2 ml of chloroform, gradually mixed with 50 mg of poly(ethylene oxide) - PEO with continuous mixing. Subsequently, the polystyrene and PEO were mixed with each other and supplemented with 8 ml of the previously obtained polyaniline - PANI suspension (protects against microbiological contamination), with mixing in order to obtain a uniform suspension of polyaniline in the solution of polymers. The process of electro-spinning was conducted in a non-uniform electric field under the voltage of 4.5 kV, inter-electrode distance of 15 cm and polymer solution outflow rate of 0.2 mL/min. Obtained nano- and microfibre random networks were transferred from the electrode into plastic carrier frames, forming final grids.

All reagents used for the synthesis of polyaniline and fabrication of nanofibers using electro-spinning were purchased from Sigma-Aldrich, St Louis, MO, USA.

## CELL CULTURE

Human HeLa cervical epithelial cells (ATCC CCL-2) were purchased from the American Type Cell Culture (ATCC, Manassas, MA). HeLa cells were cultured in standard conditions: RPMI medium (PAN Biotech-Gmbh, Aidenbach, Germany) supplemented with 10% fetal calf serum (FCS; Sigma, St Louis, MO, USA), 2 mmol/L L-glutamine (Cambrex, Charles City, IA, USA), 100 IU/mL penicillin and streptomycin solution (Sigma-Aldrich, St Louis, MO, USA) in a 5% CO2-humidified atmosphere at 37°C.

In order to examine the usefulness of nanostructural grids as a tool for 3D cultures, the cells were seeded on grids at a density of 10 000 cells. Subsequently, the grids with cells were placed into fresh culture medium and after 24, 48 or 72 hours cells morphology and their viability were examined. Every experiment was repeated ten times.

### Cell Viability Assessment

The viability of HeLa cells was analyzed using the XTT colorimetric test, based on the reduction of XTT compound (tetrazoline-2,3-bis(2-methoxy-4-nitro-5-sulphophenyl)-2H-5-carboxyanilide,Sigma-Aldrich, St Louis, MO, USA) by living cells. Optical density of colour product was measured at the wavelength of 450 nm.

The percentage of mitochondrial activity was calculated according to the following equation: (OD of nanostructured grid with cells−OD of medium alone)/(OD of nanostructured grid only−OD medium alone)×100; where OD is optical density [18]. Statistical analysis of the results was done with the Kruskal-Wallis test with Dunn's post-test using STATISTICA ver.5 software (Statsoft, Krakow, Poland). *P* value less than 0.05 was considered statistically significant.

### Microscopy

#### *Fluorescence microscopy.*

After 24, 48 or 72 hours of culture on the grids HeLa cells were stained with 0.1 μg/ml of Hoechst 33342 and 0.125 μg/ml of propidium iodide (Sigma-Aldrich, St Louis, MO, USA). The presence of intact, apoptotic and/or necrotic cells was evaluated using Nicon Diaphot Eclipse TE 200 fluorescence microscope, equipped with UV-2A and FITC/FLUO-3 filters [19].

## Confocal microscopy

In order to visualize the 3D construct the cells were pre-fixed for 10 minutes in 4% paraformaldehyde (Sigma-Aldrich, St Louis, MO, USA) and stained with 0.125 μg/ml of propidium iodide (Sigma-Aldrich, St Louis, MO, USA). Signals were excited at 543 nm wavelength and fluorescence emission was selected with 560 nm bandpass filter. Images of cells were represented as orthogonal projections of 20–60 optical sections in 0.3 μm increments using Zeiss LSM 510 confocal microscope.

## Scanning electron microscopy

After 48 hours of culture on the grids the cells were fixed in a standard way using 2.5% glutaraldehyde (at room temperature for 60 min and, then at 4°C for 24 hours). The grids with cells were washed with PBS for 30 min and the cells were gradually dehydrated in alcohols. The grids with cells were finally coated with palladium and visualized with a scanning electron microscope (Philips SEM 515).

## Transmission electron microscopy

After 48 hours the nanostructured grids with the cells were put into a fixative solution containing 4% glutaraldehyde (Taab, Berkshire, UK), buffered with 0.1 M phosphate buffer, pH 7.3 (Merck). After overnight incubation the nanostructured grids were rinsed for 60 min in 0.1 M phosphate buffer, pH 7.3, and then postfixed in 2% $OsO_4$ in 0.1 M phosphate buffer for 2 hours (Merck). The specimens were dehydrated in ethanol, block-stained with alcoholic uranyl acetate and embedded in Spurr's medium (Merck). After contrasting, the ultrastructure of HeLa cells was examined using JEOL 100 Transmission Electron Microscope.

## Results

### *Morphology of Polyaniline Fibers*

Light microscopy confirmed that the analyzed grids are transparent (Fig. 1A). Moreover, scanning electron microscopy showed that the thickness of electrospun fibers was 2000–2500 nm, and the diameter of pores was 50–300 nm (shorter dimension). The average thickness of a grid is less than 0.5 mm (Fig. 1B).

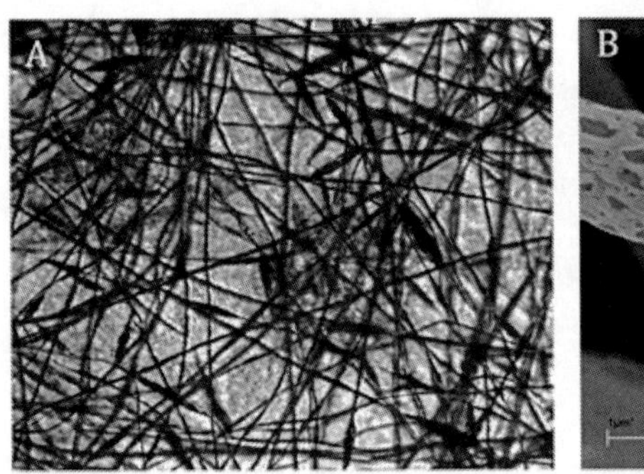


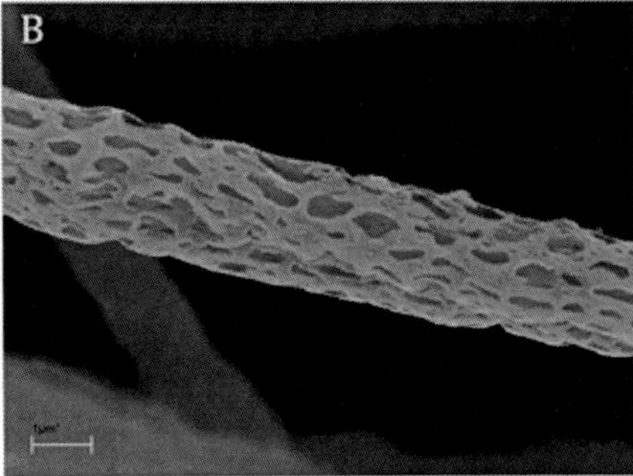


Figure 1. Light microscopy image of nanostructured grid.

Original magnification x200 (A). Scanning electron microscopy image of nanostructured microfiber. Original magnification x500 (B).

doi:10.1371/journal.pone.0072936.g001

## CELL VIABILITY

To determine whether the tested electrospun scaffolds can influence the viability of uterine cervix carcinoma cells cultured in 3D, their viability was examined using the XTT assay. Due to XTT-reducing properties of the nanostructured grids alone, the viability of cells grown on the grids were related to the results obtained for measurements of the grids alone.

As shown in Fig. 2 the grids have no cytotoxic effect on HeLa cells. After 48 hours of culture on the nanostructured grids cell viability increased compared to the evaluation made after 24 hours. However, the change was insignificant ($p>0.05$). Subsequent incubation, up to 72 hours, resulted in a significant increase of cell viability, compared to the viability noted after either 24 hour ($p<0.001$) or 48 hour of culture ($p<0.05$) (Fig. 2).

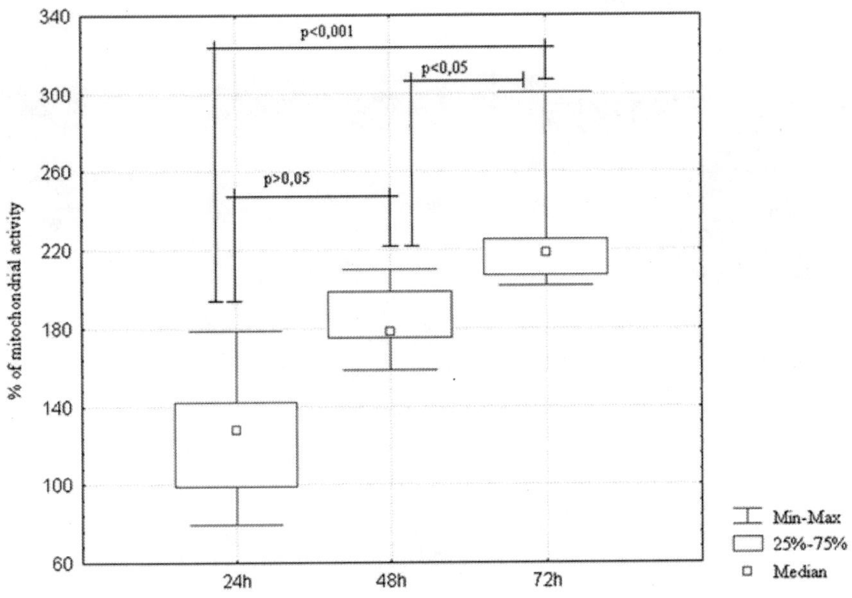


Figure 2. The viability of HeLa cells cultured on the nanostructured grid.

HeLa cells were incubated for 24, 48 72 hours on the nanostructured grids and the percentage of mitochondrial activity was calculated according to the following formula: (OD of nanostructured grid with cells−OD of medium alone)/(OD of nanostructured grid only−OD medium alone)×100; where OD is optical density.

doi:10.1371/journal.pone.0072936.g002

## Cell Morphology

The double staining using Hoechst 33342 and propidium iodide allows distinguishing living and necrotic cells as propidium iodide penetrates only cells with a disrupted cell membrane, while Hoechst

33342 penetrates also intact cell membranes. As it was shown in Fig. 3 the nuclei of cells cultured on the nanostructured grids 24 (Fig. 3A), 48 (Fig. 3B) and 72 (Fig. 3C) hours emitted bright blue fluorescence. The staining proved that in the analyzed time intervals cells showed high viability with very low level of necrosis. Moreover, confocal microscopy confirmed that the cells cultured on the grids formed a homogenous 3D constructs (Fig. 3D).

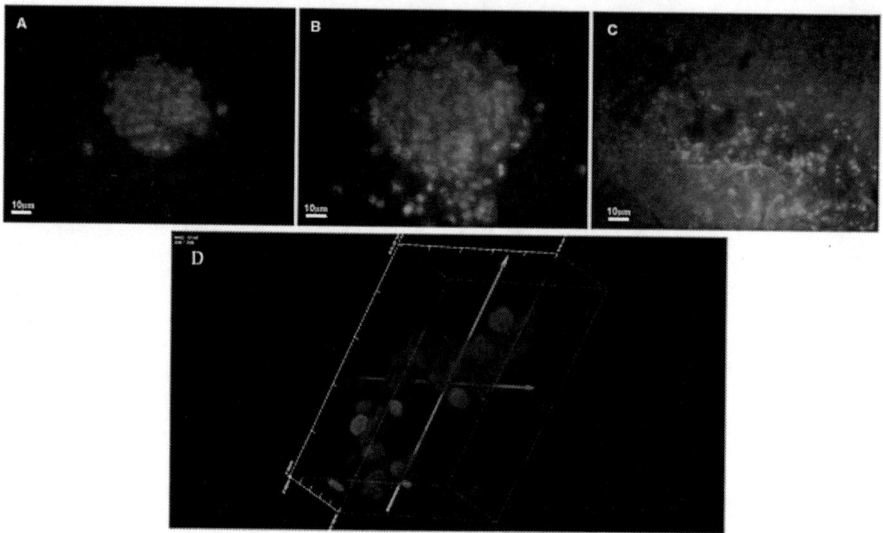


Figure 3. HeLa cells stained with Hoechst 33342 and propidium iodide after 24 (A), 48 (B) and 72 (C) hours of 3D culture on the nanostructure grids.

Original magnification x200. A fragment of the spherical 3D structure of cell growing on nanostructured grids visualized after 48 hours. Original magnification x400 (D).

doi:10.1371/journal.pone.0072936.g003

The results of the light microscopy and scanning electron microscopy observation demonstrated that after 48 hours of culture on the nanostructured grids HeLa cells formed symmetrical and spherical aggregates. The size of cells ranged between 6 and 7 micrometers (Fig. 4).

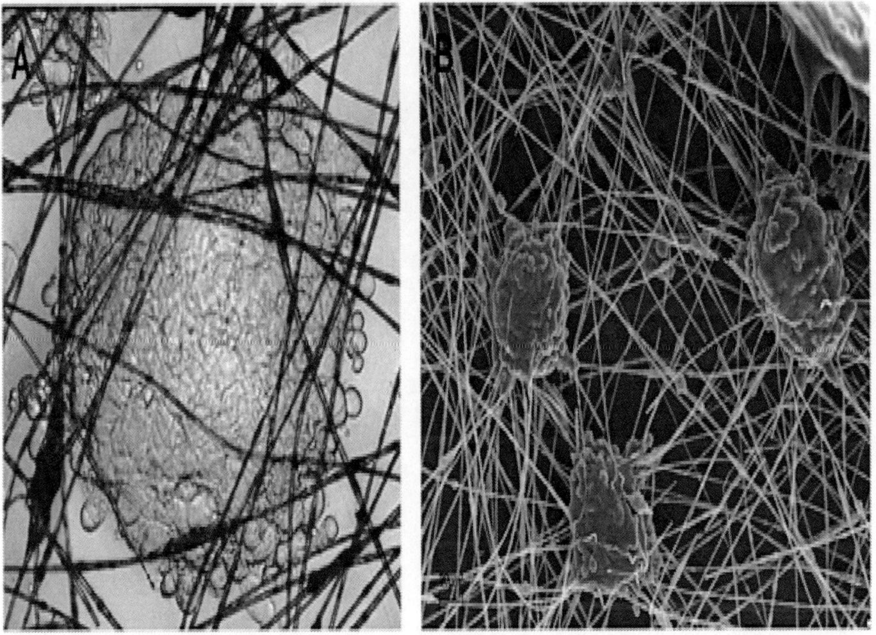


Figure 4. Cells growing on the scaffold visualized using light microscopy after 48 hours of incubation.

Original magnification x200 (A). Scanning electron microscopy image of HeLa cells on the nanostructured grid fibers after 48 hours of incubation. Original magnification x100 (B).

doi:10.1371/journal.pone.0072936.g004

Detailed analysis of the cells grown for 48 hours on the nanostructured grids showed that there were no changes in the ultrastructure of HeLa cells. The cells were oval or elongated in shape with abundant microvilli and mostly euchromatic nuclei. The cytoplasm showed numerous intact organelles. Cisterns of the Golgi apparatus frequently contained electron-dense material in the form of laminae. The cells possessed numerous mitochondria with partially obliterated inner structure, sometimes presenting cristae. Some cells contained phagolysosomes with electron-dense content (Fig. 5).

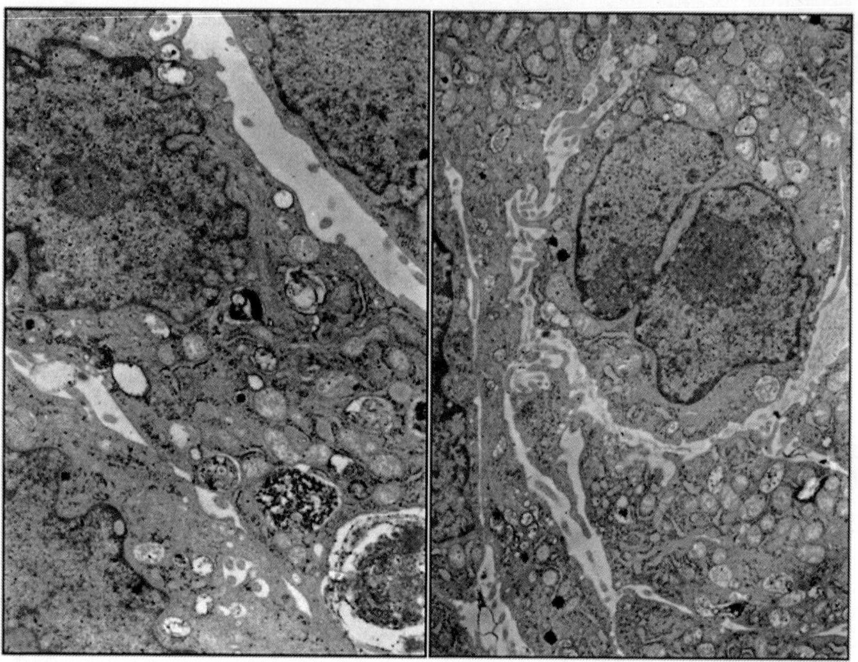

Figure 5. Transmission electron microscopy images of HeLa cells on the nanostructure grid fibers after 48 h culture.

Original magnification x6000.

doi:10.1371/journal.pone.0072936.g005

## DISCUSSION

Three-dimensional culture scaffolds assure a clearly better growth and proliferation of cells as they mimic the *in vivo* conditions. Moreover the cells cultured on 3D scaffolds have been found to develop a self-specific microenvironment [2],[13],[24]. Culture scaffolds of such properties can be used, for instance in regenerative medicine [2], [7], [9], [12], [22]–[23]. However, not all three-dimensional scaffolds are adequate for cell culture, as the building material of a scaffold must fulfill certain conditions [7], [10], [21], [25]. In particular it should allow for attachment of cells and their undisturbed growth and proliferation.

One of the most common among technologies used in production of nano- and microfibres from polymer solutions is electro-spinning. Studies confirm that this technique can be applied to obtain spatial scaffolds for cell cultures in *in vitro* conditions [20]–[21].

The electrospun three-dimensional nanostructured grids presented in this study were not only proven to be biocompatible but also suitable for cell culture. One of the advantages of these nanostructured grids is in their architecture in the form of nano- and microtubes. What is more in the process of electro-spinning these hollow fibers are being arranged randomly, forming an irregular network with micrometric distance between the tubes preserved. The electrospun capillaries possess nanopores which probably enhance the free flow of the culture medium. These features have favorable effects on cell attachment and their viability. It has also been shown that these properties may promote migration of the proliferating cells along the fibers or between the overlaying fiber layers [1].

XTT test results have demonstrated that the described scaffold is suitable for cell culture in 3D conditions. The intense and significant increase in cell mitochondrial activity was observed particularly 48 hours after seeding of cells on the grid. The observation has been additionally supported by results from fluorescent microscopy with the use of Hoechst 33342 and propidium iodide. It was demonstrated that within 24 hours of seeding, the cells attached to nanostructured grid fibers and formed spatial agglomerates, which in the course of culture formed multilayered structures.

It can be assumed that the process of cell seeding was accompanied by the selection of particular types of cells, which were capable of proliferation on the fibers and thereby accept the new supporting material. The problem which remains to be resolved involves our inability to calculate the precise number of cells deposited on the nanostructured grid in the first day of the culture. For this reason, the obtained values of mitochondrial activity have been related to results obtained for an empty nanostructured grid and then to each other.

The analysis of results obtained using a scanning electron microscope has shown that the cells deposited on nanostructured grids form protrusions, which may either anchor the cells at the site of deposition or allow their migration along the fibers. It may be expected that the anchored cells may form a barrier preventing further amoeba-like movements of the adjacent cells. This would explain the reason of the inward growth of cells and formation of the spherical spatial structures. The observed spherical shapes of cell agglomerates formed following 24 hours after cell deposition allow for local development of high density cultures on a limited area. Thus, the presented scaffold represents not only an excellent support for growing cells in 3D conditions but may also stabilize the environment of proliferating cells thanks to the specific microenvironment developed. Nevertheless, it should be noted that the above hypothesis requires verification using studies on interaction of cells with individual fibers.

The presence of pores in the described fibers not only promotes cell growth but also makes the grid transparent. The potential for cell observation and their growth monitoring using light microscope is an important advantage of the suggested method of 3D cultures. Such a feature eliminates the need for staining of cells with fluorescent markers or for their transfection using e.g. GFP. Moreover, the nanostructured grid-deposited cells may be used in studies on ultrastructure of cells in *in situ* conditions and thus the simple technique of preparing cultures for Spurr embedding may be here applied.

## CONCLUSIONS

A new solution for 3D cultures has been suggested, which is suitable for cell culture in conditions mimicking *in vivo* environment. The relatively high surface to volume ratio permits to load such a nanostructured grid with high number of cells and to grow them in conditions of high density or even overcrowding. Moreover, the suggested novel 3D support in many ways surpasses other similar culture approaches: the scaffolds are transparent and thus allow observation of viable cells in time; standard fluorescence or colorimetric methods such can be used for evaluation of cells. Finally, the cell nanostructured grid can be embedded in *in situ*conditions for ultrastructural studies.

## AUTHOR CONTRIBUTIONS

Conceived and designed the experiments: MWC KL KS AG AJ WW JL. Performed the experiments: MWC KL AG. Analyzed the data: MWC KL WW. Contributed reagents/materials/analysis tools: MWC. Wrote the paper: MWC AJ JL.

## REFERENCES


1. Yamada KM, Cukierman E (2007) Modeling tissue morphogenesis and cancer in 3D. Cell 130: 601–10 10.1016/j.cell.2007.08.006.
2. Ma PX (2008) Biomimetic materials for tissue engineering. Adv Drug Deliver Rev 60: 184–98 10.1016/j.addr.2007.08.041.
3. Fischbach C, Chen R, Matsumoto T, Schmelzle T, Brugge JS, et al. (2007) Engineering tumors with 3D scaffolds. Nat Methods 4: 855–60 10.1038/nmeth1085.
4. Jang JH, Castano O, Kim HW (2009) Electrospun materials as potential platforms for bone tissue engineering. Adv Drug Deliver Rev 61: 1065–83 10.1016/j.addr.2009.07.008.
5. Zhang S (2008) Designer self-assembling peptide nanofiber scaffolds for study of 3-D cell biology and beyond. Adv Cancer Res 99: 335–62 10.1016/S0065-230X(07)99005-3.
6. Kraehenbuehl TP, Langer R, Ferreira LS (2011) Three-dimensional biomaterials for the study of human pluripotent stem cells. Nat Methods 8: 731–6 10.1038/nmeth.1671.
7. Shin H, Jo S, Mikos AG (2003) Biomimetic materials for tissue engineering. Biomaterials 24: 4353–4364 10.1016/S0142-9612(03)00339-9.
8. Kim JH, Park CH, Lee OJ, Lee JM, Kim JW, et al. (2012) Preparation and in vivo degradation of controlled biodegradability of electrospun silk fibroin nanofiber mats. J Biomed Mater Res A 100: 3287–95 10.1002/jbm.a.34274.
9. Jeong SI, Krebs MD, Bonino CA, Khan SA, Alsberg E (2010) Electrospun alginate nanofibers with controlled cell adhesion for tissue engineering. Macromol Biosci 10(8): 934–43 10.1002/mabi.201000046.
10. Jo JH, Lee EJ, Shin DS, Kim HE, Kim HW, et al. (2009) In vitro/in vivo biocompatibility and mechanical properties of bioactive glass nanofiber and poly(epsilon-caprolactone) composite materials. J Biomed Mater Res B Appl Biomater 91(1): 213–2010.1002/jbm.b.31392.

11. Lee JJ, Yu HS, Hong SJ, Jeong I, Jang JH, et al. (2009) Nanofibrous membrane of collagen-polycaprolactone for cell growth and tissue regeneration. J Mater Sci Mater Med 20: 1927–35 10.1007/s10856-009-3743-z.

12. Cai YZ, Zhang GR, Wang LL, Jiang YZ, Ouyang HW, et al. (2012) Novel biodegradable three-dimensional macroporous scaffold using aligned electrospun nanofibrous yarns for bone tissue engineering. J Biomed Mater Res A 100: 1187–94 10.1002/jbm.a.34063.

13. Davidenko N, Campbell JJ, Thian ES, Watson CJ, Cameron RE (2010) Collagen-hyaluronic acid scaffolds for adipose tissue engineering. Acta Biomater 6: 3957–6810.1016/j.actbio.2010.05.005.

14. Ifkovits JL, Burdick JA (2007) Review: photopolymerizable and degradable biomaterials for tissue engineering applications. Tissue Eng 13: 2369–85 10.1089/ten.2007.0093.

15. Rnjak-Kovacina J, Weiss AS (2011) Increasing the pore size of electrospun scaffolds. Tissue Eng Part B Rev 17: 365–72 10.1089/ten.teb.2011.0235.

16. Nam J, Huang Y, Agarwal S, Lannutti J (2007) Improved cellular infiltration in electrospun fiber via engineered porosity. Tissue Eng 2007 13: 2249–5710.1089/ten.2006.0306.

17. Kim TG, Park TG (2006) Biomimicking extracellular matrix: cell adhesive RGD peptide modified electrospun poly(D,L-lactic-co-glycolic acid) nanofiber mesh. Tissue Eng 12: 221–33 doi:10.1089/ten.2006.12.221.

18. Wołuń-Cholewa M, Szymanowski K, Nowak-Markwitz E, Warchoł W (2010) Photodiagnosis and photodynamic therapy of endometriotic epithelial cells using 5-aminolevulinic acid and steroids. Photodiagnosis Photodyn Ther 8: 58–6310.1016/j.pdpdt.2010.12.003.

19. Fik E, Wołuń-Cholewa M, Kistowska M, Warchoł JB, Goździcka-Józefiak A (2001) Effect of lectin from Chelidonium majus L. on normal and cancer cells in culture. Folia Histochem Cytobiol 39: 215–216.

20. Ahmed I, Ponery AS, Nur-E-Kamal A, Kamal J, Meshel AS, et al. (2007) Morphology, cytoskeletal organization, and myosin dynamics of mouse embryonic fibroblasts cultured on nanofibrillar surfaces. Mol Cel Biochem 301: 241–9 10.1007/s11010-007-9417-6.

21. Rodrigues MT, Martins A, Dias IR, Viegas CA, Neves NM, et al. (2012) Synergistic effect of scaffold composition and dynamic culturing environment in multilayered systems for bone tissue engineering. J Tissue Eng Regen Med 6: e24–3010.1002/term.499.

22. Garg K, Bowlin GL (2011) Electrospinning jets and nanofibrous

structures. Biomicrofluidics 5: 13403 10.1063/1.3567097.

23. Sill TJ, Von Recum HA (2008) Electrospinning: applications in drug delivery and tissue engineering. Biomaterials 29: 1989–2006 10.1016/j.biomaterials.2008.01.011.
24. Bhattarai SR, Bhattarai N, Yi HK, Hwang PH, Cha DI, et al. (2004) Novel biodegradable electrospun membrane: scaffold for tissue engineering. Biomaterials 25: 2595–260210.1016/j.biomaterials.2003.09.043.
25. Tuzlakoglu K, Bolgen N, Salgado AJ, Gomes ME, Piskin E, et al. (2005) Nano- and micro-fiber combined scaffolds: a new architecture for bone tissue engineering. J Mater Sci Mater Med 16: 1099–104 10.1007/s10856-005-4713-8.

# Chapter 9

# INFLUENCE OF MATERIAL PROPERTIES ON TIO2 NANOPARTICLE AGGLOMERATION

Dongxu Zhou[1,2], Zhaoxia Ji[2], Xingmao Jiang[3], Darren R. Dunphy[3], Jeffrey Brinker[3,4], Arturo A. Keller[1,2*]

[1]Bren School of Environmental Science and Management, University of California Santa Barbara, Santa Barbara, California, United States of America

[2]University of California Center of Environmental Implications of Nanotechnology, University of California Los Angeles, Los Angeles, California, United States of America, [3]Center for Micro-Engineered Materials and Department of Chemical Engineering, University of New Mexico, Albuquerque, New Mexico, United States of America, [4]Sandia National Laboratory, Albuquerque, New Mexico, United States of America

## ABSTRACT

Emerging nanomaterials are being manufactured with varying particle sizes, morphologies, and crystal structures in the pursuit of achieving outstanding functional properties. These variations in these key material properties of nanoparticles may affect their environmental fate and transport. To date, few studies have investigated this important aspect of nanoparticles' environmental behavior. In this study, the aggregation kinetics of ten different $TiO_2$nanoparticles (5 anatase and 5 rutile each with varying size) was systematically evaluated. Our results show that, as particle size increases, the surface charge of both anatase and rutile $TiO_2$nanoparticles shifts toward a more negative value, and, accordingly, the point of zero charge shifts toward a lower value. The colloidal stability of anatase sphere samples agreed well with DLVO theoretical predictions, where an increase in particle size led to a higher energy barrier and therefore greater critical coagulation concentration. In contrast, the critical coagulation concentration of rutile rod samples correlated positively with the specific surface area, i.e., samples with higher specific surface area exhibited higher stability. Finally, due to the large innate negative surface charge of all the $TiO_2$ samples at the pH value (pH = 8) tested, the addition of natural organic matter was observed to have minimal effect on $TiO_2$ aggregation kinetics, except for the smallest rutile rods that showed decreased stability in the presence of natural organic matter.

## INTRODUCTION

Given the accelerating production of existing and emerging engineered nanoparticles (ENPs), the accidental spill and use-phase or end-of-product-life release of nanoparticles into the environment may be inevitable [1]–[3]. In fact, a few studies have already reported detectable levels of TiO2 nanoparticles in a wastewater treatment plant and in natural water streams[4]–[6]. To accurately assess the environmental distribution, the major sinks, and the ecological risks of ENPs, a comprehensive understanding of how ENPs behave in the aqueous environment is imperative [1], [2], [7]–[9].

The fate and transport of ENPs in the aqueous environment is controlled by both the chemistry of the aqueous systems and the material properties of the ENP [2], [10]–[12]. In recent years, the effect of solution chemistry on the aggregation of mostly spherical ENPs has been extensively studied [13]–[19] and is relatively well understood. For instance, pH alters the colloidal stability of the ENPs system by modulating the protonation/deprotonation equilibrium and further altering the electrostatic repulsion [13], [17]. Indifferent electrolytes compress the nanoparticle electric double layer and reduce the energy barrier [20], [21]. The presence of natural organic matter, depending on the concentration, can either stabilize nanoparticles by providing additional electrostatic repulsion and/or steric hindrance, or bridge multiple particles and enhance aggregation [16], [19]. Our recent study revealed that natural clay minerals can coagulate either positively or negatively charged nanoparticles due to their edge-face charge heterogeneity [22].

On the other hand, only until recently limited studies started to investigate the effect of intrinsic material properties of ENPs, such as particle size, morphology, crystal structure, and dopants on ENPs' aggregation and mobility [2], [23]. Kobayashi et al. demonstrated that an additional repulsive force appears on silica surfaces as particle size decreases [24]. He et al. showed that larger hematite nanoparticles (65 nm and 32 nm) were more stable than smaller ones (12 nm), which was qualitatively explained by DLVO theory [25]. Mulvihill et al. investigated the effects of three stabilizing agents on the colloidal stability of CdSe nanoparticles (4, 6, and 8 nm spheres and 2.9×24 nm rods) [26], and they found capping ligand dissociation to be the primary nanoparticle aggregation mechanism. In addition, the critical coagulation concentrations of four different CdSe nanoparticles were found to be linearly correlated to their specific surface area[26]. Liu et al. reported that 50 nm anatase spheres and 10×40 nm rutile rods settled more slowly than 5 and 10 nm anatase nanospheres. Differences in the amounts of sulfur and phosphate impurities introduced during the synthesis process determine the stability of $TiO_2$spheres and rods [27]. These studies suggest that material properties such as particle size, capping ligand, and impurities are important parameters affecting nanoparticles' aqueous stability. However, a systematic study on the role of intrinsic properties of nanoparticle aggregation where particle size is progressively varied and crystal structure is controlled is lacking.

The goal of this study was to investigate the influence of particle size, morphology, and crystal structure in $TiO_2$ nanoparticle aggregation. The two most abundant polymorphs of $TiO_2$ are rutile and anatase, both crystalize in the tetragonal system. Rutile is the stable phase, and anatase is metastable [28]. Rutile and anatase of increasing size were synthesized and their colloidal stability was characterized by means of electrophoretic mobility and light scattering. We present results on particle charge, critical coagulation concentrations, and absolute doublet formation rates. We found that no single material property was a determining factor that controls $TiO_2$ aggregation; rather, a combination of various material parameters needs to be considered to predict nanoparticle aggregation.

## MATERIALS AND METHODS

### Materials

Ten $TiO_2$ samples, five rutile rods (designated RR) and five anatase spheres (designated AS), with varying sizes were synthesized via a hydrothermal approach. In a typical synthesis of spherical anatase NPs (AS samples), 1.34 g of amorphous titanium dioxide (NanoActive, Nanoscale Corp.) was added to 43 g of 1 M $H_2SO_4$ and heated to 230°C for 24 hours in a Parr bomb. For rutile NPs rods (RR samples), 23.7 g of $TiCl_4$ was dissolved in 50 ml of 37% HCl; 6.0 g of this solution was added to 11.7 g of 1 M HCl and then heated to 200°C for 24 hours, again in a Parr bomb. NP size for both rutile and anatase was varied by control of the precursor/acid ratio, with increased precursor concentration yielding progressively larger particle sizes. To remove residual salt, sample suspensions were dialyzed (MWCO 12–14k, Spectrum Laboratories, CA) against de-ionized water until the conductivity inside and outside of the membrane were identical.

Sample crystal structure was characterized by X-ray Diffraction (XRD) (X'pert Powder, PANalytical, the Netherlands). Particle morphology and size were assessed by transmission electron microscopy at 80 kV (JEOL 1230, JEOL, Japan). TEM samples were prepared by placing a drop of the $TiO_2$ suspension on a 200 mesh copper grid (Ted Pella, CA) and allowing it to air dry overnight. ImageJ software (NIH, USA) was used to determine the particle dimensions. The number-weighted dimensions (diameter for spheroids; length × width for rods) are determined by measuring 100 randomly selected nanoparticles on the TEM images, and the surface areas are calculated assuming a sphere shape for anatase samples and a cylinder shape (length as the cylinder height and width as the base diameter) for rutile samples (Table 1). For comparison purposes, all RR samples had an aspect ratio of around 3.5–4.5. The RR samples were relatively monodisperse, with a coefficient of variation (CV) in the range of 0.2–0.35, except RR4 which has a much higher polydispersity (CV≈0.6). The number-weighted diameter of AS particles ranged between 6–150 nm. The CV values were 0.2–0.3.

| Sample | Primary Particle Dimension (nm)[a] | Specific surface area ($m^2/g$) [b] | Crystal Struture | CCC (mM), without NOM | CCC (mM), with 10 mg/L NOM |
|---|---|---|---|---|---|
| RR1 | $15\pm5\times4\pm1$ | 236.4 | rutile | 240 | 80 |
| RR2 | $41\pm14\times10\pm2$ | 78.8 | rutile | 77 | 65 |
| RR3 | $121\pm33\times29\pm7$ | 31.5 | rutile | 65 | -[c] |
| RR4 | $201\pm122\times40\pm32$ | 23.6 | rutile | 75 | 40 |
| RR5 | $193\pm64\times52\pm12$ | 18.2 | rutile | 25 | 30 |
| AS1 | $6\pm2$ | 59.1 | anatase | 25 | 40 |
| AS2 | $11\pm3$ | 32.2 | mixture of rutile and anatase | 25 | 15 |
| AS3 | $38\pm7$ | 9.3 | anatase | 30 | 50 |
| AS4 | $54\pm17$ | 6.6 | anatase | 65 | 50 |
| AS5 | $152\pm43$ | 2.3 | anatase | 100 | 25 |

[a]primary particle dimensions are determined by measuring 100 randomly-chosen particles on TEM images.
[b]specific surface area is calculated based on the particle dimensions, assuming rutile rods as cylinders and anatase spheres as perfect spheres. The density of TiO2 was $4.23\times10^6$ g/m$^3$.
[c]CCC of RR3 with NOM appeared to be below the lowest tested electrolyte concentration, thus no CCC is reported here.
doi:10.1371/journal.pone.0081239.t001

Table 1. Measured properties of the $TiO_2$ samples.

doi:10.1371/journal.pone.0081239.t001

Suwannee River natural organic matter (NOM) was purchased from the International Humic Substance Society (IHSS, GA, USA). A 200 mg/L stock was prepared by dissolving NOM in deionized water. All reagents used in this study were analytical grade. NaCl (Sigma-Aldrich) was used as the indifferent electrolyte. Borate buffer was used to maintain a constant pH = 8.0. HCl (0.1 M and 0.01 M, EMD Chemicals Inc.) and NaOH (0.1 M, Fisher Scientific) were used as titrants for point of zero charge titration. All the solutions were filtered via 0.22 mm PVDF filters to avoid potential interference in the light scattering experiments.

## ELECTROPHORETIC MOBILITY MEASUREMENTS

Electrophoretic mobility was measured on a Malvern Zetasizer coupled with a MPT-2 autotitrator (Malvern, UK). 0.1 M, 0.01 M HCl and 0.1 M NaOH were used as titrates. A 120 s equilibribration time was allowed before each measurement. At each pH value, data were collected in triplicate.

## AGGREGATION KINETICS

The detailed procedure for measuring aggregation kinetics is described elsewhere [22]. Briefly, predetermined amounts of buffer stock (20 mM), NaCl stock (1 M, 0.1 M, and 0.01 M), NOM (1 g/L, when the effect of NOM was investigated), and deionized water were mixed together to make up 0.9 mL total volume at the desired pH and ionic strength.

This mixture was added into 0.1 mL of a specific $TiO_2$ stock suspension immediately before the aggregation kinetics measurements. The hydrodynamic diameter of the suspension was monitored by dynamic light scattering (Malvern Zetasizer Nano, UK) as a function of time. The detection angle was 90°, and the laser wavelength was 633 nm. The cumulant algorithm was used to calculate the hydrodynamic diameter. The measurements lasted until either the hydrodynamic diameter of the sample doubled or the measurement duration reached 1 hr, whichever criterion was met first.

The slope of the early-stage aggregation (arbitrarily fixed as the time from $a_0$ to $1.5\times a_0$) curve was then used to determine the doublet formation rate according to [14]: $\left(\frac{da_h(t)}{dt}\right)_{t\to 0} \propto k_{11} N_0$ here $a_h(t)$ is the hydrodynamic diameter of agglomerates as a function of time $t$, $N_0$ is the initial number concentration of primary particles, and $k_{11}$ is the doublet formation rate. The attachment efficiency (α) - salt concentration plot is typically characterized by two regimes, the reaction- and diffusion-limited cluster aggregation regimes (RLCA and DLCA). α increases with salt concentration in the RLCA regime; while it is independent to salt concentration in the DLCA regime. The turning point between the two regimes is called the critical coagulation concentration (CCC).

# RESULTS AND DISCUSSION

## Nanoparticle characterization

Morphology and surface charge characteristics of the $TiO_2$ samples were quantified by TEM (Figure 1 and File S1) and electrophoresis, respectively. Regardless of the morphology, crystal structure, and size, all $TiO_2$ samples showed typical amphoteric charging patterns [29]. Three types of groups (singly coordinated $Ti_3O^0$, doubly coordinated $Ti_2O^{2/3-}$, and triply coordinated $TiO^{4/3-}$) with varying $H^+/OH^-$ affinity constants (pK) exist on the $TiO_2$ surface [30]. Due to the extremely low pK value (−7.5) for $TiO^{4/3-}$, the singly and doubly coordinated groups determine the actual $TiO_2$ surface charge. The point of zero charge (PZC) values ranged from pH 3 to 6. These results are in good agreement with previously published values [13], [27], [30]–[32]. A correlation between the electrophoretic mobility and particle size was observed for both anatase spheres and rutile rods (Figure 2).

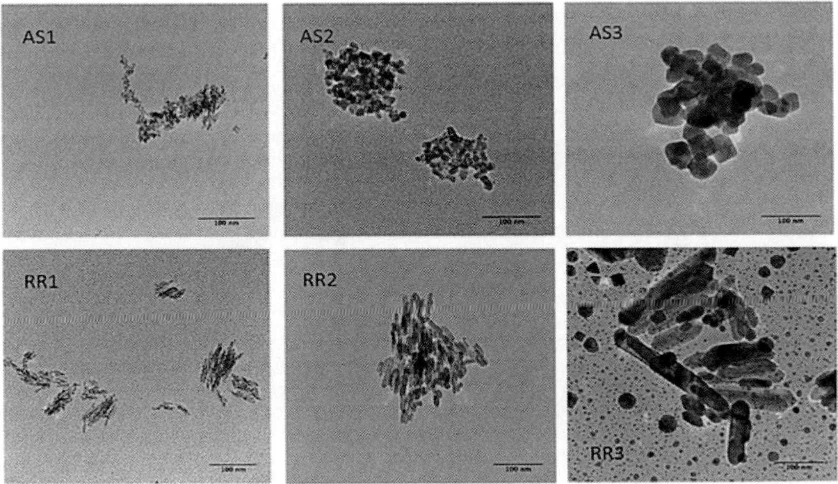


Figure 1. Representative TEM images of rutile rods and anatase spheroids.

doi:10.1371/journal.pone.0081239.g001

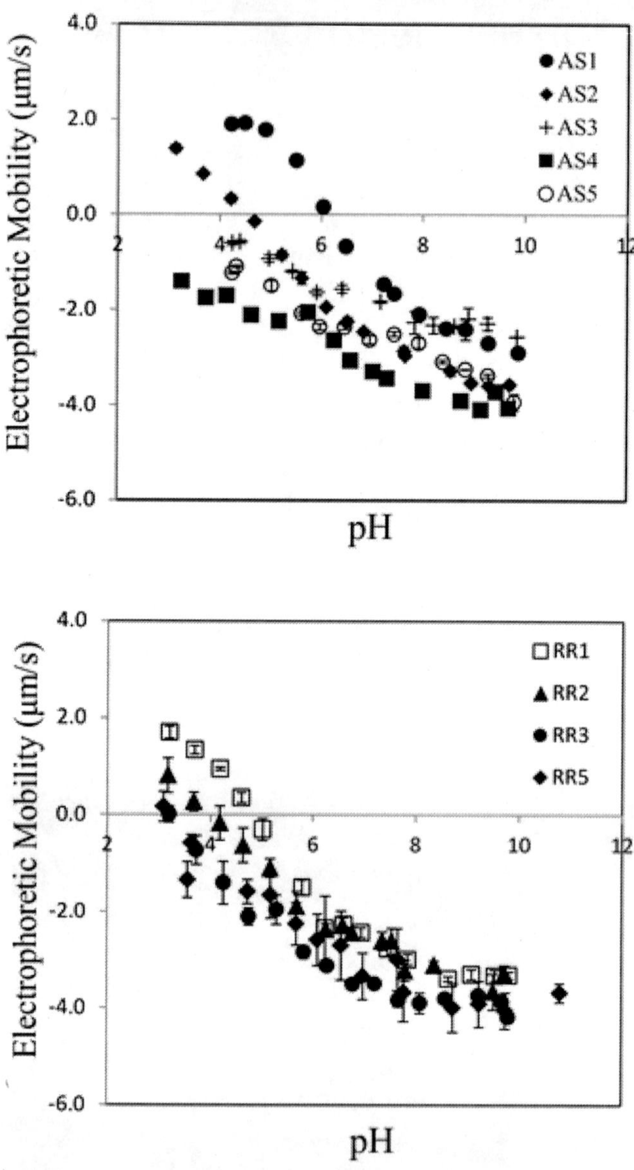


Figure 2. Electrophoretic moblity of anatase spheres (a) and rutile rods (b) as a function of pH.

A general trend of PZC shift toward a lower pH can be observed for both anatase spheres and rutile rods.

doi:10.1371/journal.pone.0081239.g002

The point of zero charge (PZC) shifts toward a lower pH value with increasing particle size for the same aspect ratio. This trend was observed experimentally for $TiO_2$ anatase spheres by others [27], [33]. Using a corrected Debye-Huckel theory and Monte Carlo simulation, Abbas et al. showed theoretically that such a size-dependence of surface charge exists for metal oxide nanoparticles [34]. It is suggested that, as particle size decreases, the curvature of nanoparticles approaches the same length scale as the hydrated ions, which enables the counterions to screen the surface sites from all directions rather than only one-half of the space in the case of a planar wall.

## AGGREGATION KINETICS

The aggregation kinetics of the various $TiO_2$ nanoparticles was examined over a wide range of NaCl concentrations (1–500 mM). A representative aggregation kinetics curve is shown in Figure S2 in File S1 and an attachment efficiency - electrolyte concentration plot is shown inFigure 3 (to avoid redundancy, additional aggregation kineticscurvesarenotshown).Thereactionlimitedclusteraggregation (RLCA) and diffusion limited cluster aggregation (DLCA) regimes can be identified in the stability plots of every $TiO_2$ sample. It appears that the electrostatic and van der Waals interactions control the $TiO_2$ aggregation process even for diverse morphologies [13], [18], [27], [29]. At low NaCl concentrations, electrostatic repulsion dominates the inter-particle interaction and aggregation occurs at a relatively slow rate. As electrolyte concentration increases, the electrostatic repulsion is suppressed due to the electric double layer (EDL) compression, and the aggregation is accelerated. Once the electrolyte concentration is high enough to completely eliminate the energy barrier, van der Waals attraction starts to dominate and most collisions between $TiO_2$ nanoparticles lead to attachment. This electrolyte concentration, called critical coagulation concentration (CCC), can serve as an index to compare nanoparticle suspension stability. The CCC values are summarized in Table 1.

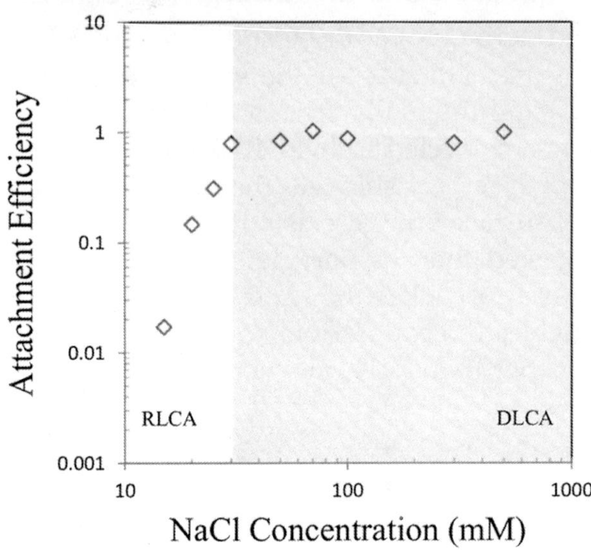


Figure 3. The attachment efficiency as a function of NaCl concentration for AS3 (anatase sphere).

A reaction-limited cluster aggregation regime (RLCA, left region) and a diffusion-limited cluster aggregation regime (DLCA, right region) can be observed.

doi:10.1371/journal.pone.0081239.g003

## INFLUENCE OF MATERIAL PROPERTIES

Under the same solution chemistry, particle size has a clear effect on the stability of the $TiO_2$AS samples (Table 1). Smaller particles are much more prone to agglomerate; an ionic strength that is typical for surface water or groundwater [3] can completely destabilize the suspension. Larger AS particles exhibit larger CCC values. A linear correlation was found between particle size and CCC, with a $R^2$ of 0.9429 (Figure 4a). A similar trend was reported for 12, 32, and 65 nm hematite nanoparticles [25]. According to DLVO theory, both van der Waals attraction and the electrostatic repulsion are functions of particle diameter (see Equation S1 and S2 in File S1). Therefore, a higher energy barrier is predicted as particle diameter increases.

Figure 4b shows the contour map of energy barrier as a function of both particle size and the ionic strength. Assuming that a suspension is completely destabilized when the energy barrier is comparable to the thermo-kinetic energy (1 kT), indeed the CCC is predicted to increase as particle size increases (Figure 4b).

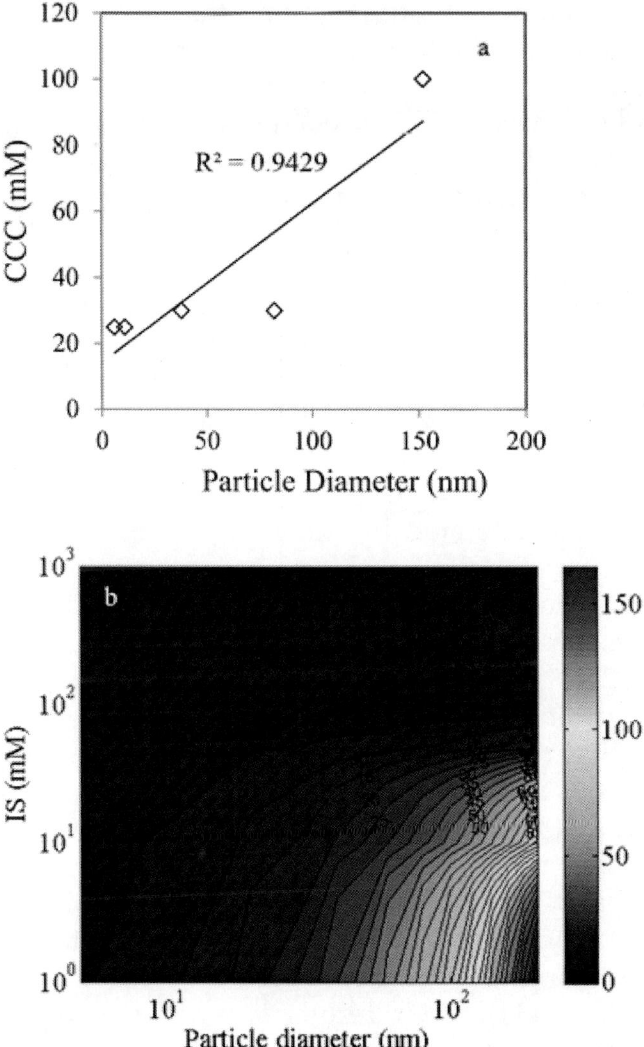


Figure 4. Anatase sphere CCC-particle size correlation and the theoretical prediction of the energy barrier.

(a). Correlation between nanoparticle diameter and CCC for anatase sphere (AS) $TiO_2$; (b). Predicted energy barrier contour map of $TiO_2$ anatase nanospheres. The color bar denotes the energy barrier (unit kT).

doi:10.1371/journal.pone.0081239.g004

The same relationship does not hold for the rutile rods samples (Table 1, Figure S3 in File S1). The 15×4 nm rutile rods are the most stable among the five rods, with a CCC c.a. 200 mM NaCl. In contrast, only 25 mM NaCl is needed to induce DLCA aggregation for RR5, the largest rutile sample. A plot of the specific surface area against CCC revealed that a proportionality exists between the two parameters for rutile rods (Figure 5). A specific surface area - CCC correlation was reported previously for CdSe nanospheres and CdSe nanorods [26]. Qualitatively the DLVO theory predicts a linear proportionality between particle size and the energy barrier, provided all the other parameters identical (Equation S1 and S2). The observed opposite trend for the rutile rods samples suggests that bulk or surface properties other than the particle size may play a central role in the stability of rutile rods. As suggested by Onsager decades ago and reiterated by McBride and Bayeye recently, for clay suspensions, particle geometry has a strong influence on the colloidal properties due to the Covolume Effect [35]. The XRD spectra revealed a systematic increase in peak intensity along the crystal face [101], [111], [211] directions as the dimension of rutile rods increased (Figure 6, remaining XRD data presented in Figure S4 in File S1. This indicates a shift in exposed crystal face composition as rutile rod size changes. Various rutile crystal faces are known to possess different surface energies [36], [37], therefore changes in exposed crystal face composition may lead to altered surface energy and in turn influence colloidal stability. In addition, Abbas and her coworkers analyzed the effect of particle size on surface charge density for metal oxide nanoparticles using the corrected Debye-Huckel theory and Monte Carlo simulation [34]. Their results revealed that a considerable increase in surface charge density occurs when the particle diameter decreases. Such an increase in surface charge density may attribute to increased stability for smaller rutile rods.

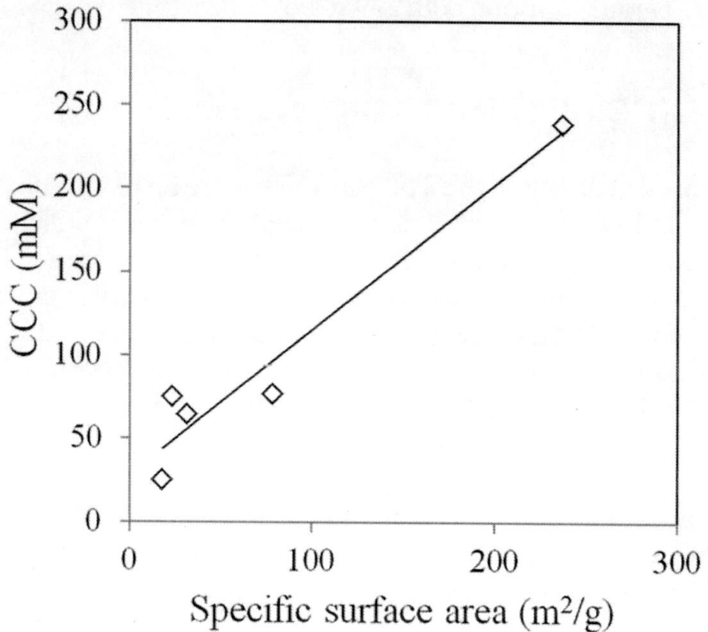


Figure 5. Correlation between specific surface area and CCC for rutile rods.

doi:10.1371/journal.pone.0081239.g005

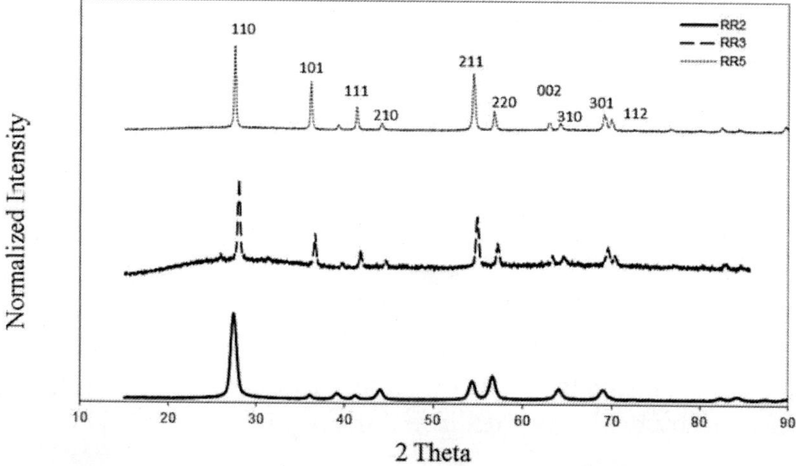


Figure 6. XRD spectra of RR2, RR3, RR4 (rutile rod) samples.

doi:10.1371/journal.pone.0081239.g006

## INFLUENCE OF HUMIC ACID

The colloidal stability of the $TiO_2$ samples were tested in the presence of 10 mg/L Suwannee River natural organic matter (NOM), and the CCC values are reported in Table 1. In our previous study, 10 mg/L organic matter (humic acid) was found to significantly enhance the stability of the P25 $TiO_2$ NPs [18]. In contrast, for most $TiO_2$ samples in this study (except RR1), the effect of NOM in altering the CCC was minor (Table 1). Chen et al. showed that the adsorption of humic acid to $TiO_2$ nanoparticles is pH dependent. At pH 5.7, humic acid significantly enhances the transport of $TiO_2$; while at pH 9.0, only a small amount of humic acid adsorbs to the $TiO_2$ surface [38]. Therefore, at the pH condition tested in this study (pH = 8), humic acid adsorption is likely to be small. Moreover, given the relatively large surface potentials of most of the $TiO_2$ samples at pH 8 (Figure 2), even the adsorbed humic acid probably only led to minimal increase of surface charge and therefore minimal change in the CCC. Thus the potential for steric interference from adsorbed NOM was not very significant for most of these particles.

For RR1, the addition of 10 mg/L NOM actually shifted the CCC towards a much lower value. Given the large specific surface area of RR1 (236.4 $m^2/g$, at least a factor of 3 larger than the rest of the samples), NOM probably could only partially cover the RR1 particles, or coat some particles and leave the remaining uncovered. In either case, there are available surfaces for a single humic acid macromolecule to attach to multiple nanoparticles; therefore the "bridging effect" may take place and facilitate aggregation [10], [20], [29]. The "bridging effect" of divalent cations is well-studied and -documented [39], [40]. In contrast, due to the low valence, the binding between monovalent cations and macromolecules are believed to be too weak to facilitate bridging. However, recent studies have shown both theoretically and experimentally that monovalent cations can indeed accelerate aggregation by bridging multiple nanoparticles[41], [42].

# CONCLUSION

We report here the distinctly different aggregation behaviors of a set of $TiO_2$ nanoparticles with varying size, crystal structure, and morphology. The isoelectric points of both anatase spheres and rutile rods shift towards a lower pH value as the particle size increases. The *CCCs* of anatase spheres correlate well with particle size, which agrees with the DLVO prediction. In contrast, *CCCs* of rutile rods exhibit a strong dependence on the specific surface area, indicating it is the surface chemistry rather than the bulk properties that dominates rutile rods aggregation. Under the conditions tested, the effect of NOM in stabilizing most $TiO_2$ samples was minor, since all the $TiO_2$ samples already possessed a large negative charge without the presence of NOM.

Nanomaterials are increasingly engineered with varying properties in the pursuit of promising biological, medical, electronic, or environmental applications. Yet the effect of nanoparticles material property alteration in controlling their environmental fate and transport is largely under-investigated. This study has shown that nanomaterials with the same chemical composition, the same solution chemistry, but different size, shape, and crystal structure have different stability and mobility. Our results stress the need to accurately characterize the material properties, such as particle size, crystal structure, and specific surface area, for a reliable prediction of the aggregation behavior of nanoparticles.

## Acknowledgments

The authors would like to thank the lab assistance of Stephen Tjan, Adeel Lakhani, Ryan Tjan, and Rachel Ker.

Disclaimer: Any opinions, findings, and conclusions or recommendations expressed in this material are those of the author(s) and do not necessarily reflect the views of the funding agencies, NSF or USEPA. This work has not been subjected to USEPA review and no official endorsement should be inferred.

## Author Contributions

Conceived and designed the experiments: DZ AAK. Performed the experiments: DZ. Analyzed the data: DZ AAK. Contributed reagents/materials/analysis tools: ZJ XJ DRD JB. Wrote the paper: DZ ZJ XJ DRD JB AAK.

## REFERENCES

1. Wiesner MR, Lowry GV, Alvarez P, Dionysiou D, Biswas P (2006) Assessing the risks of manufactured nanomaterials. Environ Sci Technol 40: 4336–4345. doi: 10.1021/es062726m
2. Hotze EM, Phenrat T, Lowry GV (2010) Nanoparticle aggregation: challenges to understanding transport and reactivity in the environment. J Environ Qual 39: 1909–1924. doi: 10.2134/jeq2009.0462
3. Keller AA, McFerran S, Lazareva A, Suh S (2013) Global life-cycle emissions of engineered nanomaterials. J Nanoparticle Res 15: 1692. doi: 10.1007/s11051-013-1692-4
4. Kaegi R, Ulrich A, Sinnet B, Vonbank R, Wichser A, et al. (2008) Synthetic TiO2 nanoparticle emission from exterior facades into the aquatic environment. Environ Pollut 156: 233–239. doi: 10.1016/j.envpol.2008.08.004
5. Kiser MA, Westerhoff P, Benn T, Wang Y, Pérez-Rivera J, et al. (2009) Titanium Nanomaterial Removal and Release from Wastewater Treatment Plants. Environ Sci Technol 43: 6757–6763. doi: 10.1021/es901102n
6. Johnson AC, Bowes MJ, Crossley A, Jarvie HP, Jurkschat K, et al. (2011) An assessment of the fate, behaviour and environmental risk associated with sunscreen TiO2 nanoparticles in UK field scenarios. Sci Total Environ 409: 2503–2510. doi: 10.1016/j.scitotenv.2011.03.040
7. Christian P, Kammer F, Baalousha M, Hofmann T (2008) Nanoparticles: structure, properties, preparation and behaviour in environmental media. Ecotoxicology 17: 326–343. doi: 10.1007/s10646-008-0213-1
8. Klaine SJ, Alvarez PJJ, Batley GE, Fernandes TF, Handy RD, et al. (2008) Nanomaterials in the environment: behavior, fate, bioavailability, and effects. Environ Toxicol Chem 27: 1825–1851. doi: 10.1897/08-090.1
9. Petosa AR, Jaisi DP, Quevedo IR, Elimelech M, Tufenkji N (2010) Aggregation and deposition of engineered nanomaterials in aquatic environments: role of physicochemical interactions. Environ Sci

Technol 44: 6532–6549. doi: 10.1021/es100598h

10. Zhou D, Keller AA (2010) Role of morphology in the aggregation kinetics of ZnO nanoparticles. Water Res 44: 2948–2956. doi: 10.1016/j.watres.2010.02.025
11. Stone V, Nowack B, Baun A, van den Brink N, Kammer FVD, et al. (2010) Nanomaterials for environmental studies: classification, reference material issues, and strategies for physico-chemical characterisation. Sci Total Environ 408: 1745–1754. doi: 10.1016/j.scitotenv.2009.10.035
12. Keller AA, Wang H, Zhou D, Lenihan HS, Cherr G (2010) Stability and aggregation of metal oxide nanoparticles in natural aqueous matrices. Environ Sci Technol 44: 1962–1967. doi: 10.1021/es902987d
13. Guzman K, Finnegan M, Banfield J (2006) Influence of surface potential on aggregation and transport of titanium dioxide nanoparticles. Environ Sci Technol 40: 7688–7693. doi: 10.1021/es060847g
14. Chen KL, Elimelech M (2007) Influence of humic acid on the aggregation kinetics of fullerene (C60) nanoparticles in monovalent and divalent electrolyte solutions. J Colloid Interf Sci 309: 126–134. doi: 10.1016/j.jcis.2007.01.074
15. Saleh NB, Pfefferle LD, Elimelech M (2008) Aggregation kinetics of multiwalled carbon nanotubes in aquatic systems: measurements and environmental implications. Environ Sci Technol 42: 7963–7969. doi: 10.1021/es801251c
16. Domingos RF, Tufenkji N, Wilkinson, K J (2009) Aggregation of titanium dioxide nanoparticles: role of a fulvic acid. Environ Sci Technol 43: 1282–1286. doi: 10.1021/es8023594
17. French RA, Jacobson AR, Kim B, Isley SL, Penn RL, et al. (2009) Influence of ionic strength, pH, and cation valence on aggregation kinetics of titanium dioxide nanoparticles. Environ Sci Technol 43: 1354–1359. doi: 10.1021/es802628n
18. Thio BJR, Zhou D, Keller AA (2010) Influence of natural organic matter on the aggregation and deposition of titanium dioxide nanoparticles. J Hazard Mater 189: 556–563. doi: 10.1016/j.jhazmat.2011.02.072
19. Thio BJR, Montes MO, Mahmoud MA, Lee DW, Zhou D, et al. (2012) Mobility of Capped Silver Nanoparticles under Environmentally Relevant Conditions. Environ Sci Technol 46: 6985–6991. doi: 10.1021/es203596w
20. Chen K, Mylon S, Elimelech M (2006) Aggregation kinetics of alginate-coated hematite nanoparticles in monovalent and divalent electrolytes. Environ Sci Technol 40: 1516–1523. doi: 10.1021/es0518068
21. Buettner KM, Rinciog CI, Mylon SE (2010) Aggregation kinetics of

cerium oxide nanoparticles in monovalent and divalent electrolytes. Colloids Surf A 366: 74–79. doi: 10.1016/j.colsurfa.2010.05.024

22. Zhou D, Abdel-Fattah AI, Keller AA (2012) Clay particles destabilize engineered nanoparticles in aqueous environments. Environ Sci Technol 46: 7520–7526. doi: 10.1021/es3004427
23. Lead J, Smith E (2009) Environmental and human health impacts of nanotechnology; Wiley.
24. Kobayashi M, Juillerat F, Galletto P, Bowen P, Borkovec M (2005) Aggregation and charging of colloidal silica particles: effect of particle size. Langmuir 21: 5761–5769. doi: 10.1021/la046829z
25. He YT, Wan J, Tokunaga T (2007) Kinetic stability of hematite nanoparticles: the effect of particle sizes. J Nanoparticle Res 10: 321–332. doi: 10.1007/s11051-007-9255-1
26. Mulvihill MJ, Habas SE, Jen-La Plante I, Wan J, Mokari T (2010) Influence of Size, Shape, and Surface Coating on the Stability of Aqueous Suspensions of CdSe Nanoparticles. Chem Mater 22: 5251–5257. doi: 10.1021/cm101262s
27. Liu X, Chen G, Su C (2011) Effects of material properties on sedimentation and aggregation of titanium dioxide nanoparticles of anatase and rutile in the aqueous phase. J Colloid Interf Sci 363: 84–91. doi: 10.1016/j.jcis.2011.06.085
28. Hanaor DAH, Sorrell CC (2011) Review of the anatase to rutile phase transformation. J Mater Sci 46: 855–874. doi: 10.1007/s10853-010-5113-0
29. Elimelech M, Gregory J, Jia X, Williams R (1995) Particle deposition and aggregation: measurement, modelling and simulation. Elsevier.
30. Giacomelli CE, Avena MJ, De Pauli CP (1995) Aspartic acid adsorption onto TiO2 particles surface. Experimental data and model calculations. Langmuir 11: 3483–3490. doi: 10.1021/la00009a034
31. Foissy A, Mpandou A, Lamarche J, Jaffrezicrenault N (1982) Surface and diffuse-layer charge at the TiO2-electrolyte interface. Colloids and surfaces 5: 363–368. doi: 10.1016/0166-6622(82)80046-2
32. Jiang J, Oberdörster G, Biswas P (2008) Characterization of size, surface charge, and agglomeration state of nanoparticle dispersions for toxicological studies. J Nanoparticle Res 11: 77–89. doi: 10.1007/s11051-008-9446-4
33. Suttiponparnit K, Jiang J, Sahu M, Suvachittanont S, Charinpanitkul T, et al. (2010) Role of Surface Area, Primary Particle Size, and Crystal Phase on Titanium Dioxide Nanoparticle Dispersion Properties. Nanoscale Res Lett 6: 27. doi: 10.1007/s11671-010-9772-1

34. Abbas Z, Labbez C, Nordholm S, Ahlberg E (2008) Size-Dependent Surface Charging of Nanoparticles. J Phys Chem C 112: 5715–5723. doi: 10.1021/jp709667u

35. McBride MB, Baveye P (2002) Diffuse double-layer models, long-range forces, and ordering in clay colloids. Soil Sci Soc Am J 66: 1207–1217. doi: 10.2136/sssaj2002.1207

36. Ramamoorthy M, Vanderbilt D, King-Smith R (1994) First-principles calculations of the energetics of stoichiometric TiO2 surfaces. Phys Rev B 49: 16721–16727. doi: 10.1103/physrevb.49.16721

37. Perron H, Domain C, Roques J, Drot R, Simoni E, et al. (2007) Optimisation of accurate rutile TiO2 (110), (100), (101) and (001) surface models from periodic DFT calculations. Theor Chem Acc 117: 565–574. doi: 10.1007/s00214-006-0189-y

38. Chen G, Liu X, Su C (2012) Distinct Effects of humic acid on transport and retention of TiO2 rutile nanoparticles in saturated sand columns. Environ Sci Technol 46: 7142–7150. doi: 10.1021/es204010g

39. Liu X, Wazne M, Han Y, Christodoulatos C, Jasinkiewicz K (2010) Effects of natural organic matter on aggregation kinetics of boron nanoparticles in monovalent and divalent electrolytes. J Colloid Interf Sci 348: 101–107. doi: 10.1016/j.jcis.2010.04.036

40. Abe T, Kobayashi S, Kobayashi M (2011) Aggregation of colloidal silica particles in the presence of fulvic acid, humic acid, or alginate: effects of ionic composition. Colloid Surface A 379: 21–26. doi: 10.1016/j.colsurfa.2010.11.052

41. Wang DW, Tejerina B, Lagzi I, Kowalczyk B, Grzybowski BA (2010) Bridging interactions and selective nanoparticle aggregation mediated by monovalent cations. ACS nano 5: 530–536. doi: 10.1021/nn1025252

42. Guerrero-Garcia GI, Gonzalez-Mozuelos P, de la Cruz MO (2011) Potential of mean force between identical charged nanoparticles immersed in a size-asymmetric monovalent electrolyte. J Chem Phys 135: 164705. doi: 10.1063/1.3656763

# Chapter 10

# A TEMPLATE-FREE, ULTRA-ADSORBING, HIGH SURFACE AREA CARBONATE NANOSTRUCTURE

Johan Forsgren[1*], Sara Frykstrand[1], Kathryn Grandfield[2], Albert Mihranyan[1*], Maria Strømme[1*]

[1]Division for Nanotechnology and Functional Materials, Department of Engineering Sciences, The A ngstrom Laboratory, Uppsala University, Uppsala, Sweden

[2]Division for Applied Materials Science, Department of Engineering Sciences, The A˚ngstro¨m Laboratory, Uppsala University, Uppsala, Sweden

## ABSTRACT

We report the template-free, low-temperature synthesis of a stable, amorphous, and anhydrous magnesium carbonate nanostructure with pore sizes below 6 nm and a specific surface area of ~ 800 $m^2$ $g^{-1}$, substantially surpassing the surface area of all previously described alkali earth metal carbonates. The moisture sorption of the novel nanostructure is featured by a unique set of properties including an adsorption capacity ~50% larger than that of the hygroscopic zeolite-Y at low relative humidities and with the ability to retain more than 75% of the adsorbed water when the humidity is decreased from 95% to 5% at room temperature. These properties can be regenerated by heat treatment at temperatures below 100°C.The structure is foreseen to become useful in applications such as humidity control, as industrial adsorbents and filters, in drug delivery and catalysis.

## INTRODUCTION

Nanotechnology is starting to influence most scientific areas and this key enabling technology is foreseen to significantly impact all materials science dependent industries during the coming decades [1]–[4]. The interest in high surface area nanostructured materials from 1990 onwards has increased exponentially for all classes of porous materials and at the beginning of 2013, according to the ISI Web of Knowledge, there were in total about 60,500 records on zeolites, 20,500 records for mesoporous silica and 13,100 records on metal organic framework (MOF) materials, whereas before 1990 these numbers were insignificant. The most common way to produce high surface area materials with micro-mesoporous structures, i.e., pores with diameters below 50 nm, is by using soft templates and building around them a more rigid structure after which the template is eluted with a solvent or burnt away to produce the rigid porous material.

In the current work we will show that it is possible, at low temperatures and without the use of templates, to synthesize a unique high surface area nanostructure with a well-defined pore-size distribution of sub 6 nm pores of a widely used, non-toxic and GRAS (generally-recognised-as-safe)-listed material that is already included in the FDA Inactive Ingredients Database [5]; viz. magnesium carbonate. Magnesium is the eighth most abundant element in the earth's crust and essential to most living species. It can form several structures of hydrated carbonates such as nesquehonite ($MgCO_3 \cdot 3H_2O$), and lansfordite ($MgCO_3 \cdot 5H_2O$), a number of basic carbonates such as hydromagnesite ($4\ MgCO_3 \cdot Mg(OH)_2 \cdot 4\ H_2O$), and dypingite ($4\ MgCO_3 \cdot Mg(OH)_2 \cdot 5\ H_2O$), as well as the anhydrous and rarely encountered magnesite ($MgCO_3$) [5]. In contrast to other alkali earth metal carbonates, chemists have found anhydrous magnesium carbonate difficult to produce, particularly at low temperatures. Above 100°C, magnesite (crystalline $MgCO_3$) can be obtained from $Mg(HCO_3)_2$ solutions by precipitation. However, at lower temperatures, hydrated magnesium carbonates tend to form, giving rise to what has been referred to as "*the magnesite problem*" [6].Yet, not only chemists have been intrigued by magnesium carbonates. Although abundant in nature, where crystalline forms exist as traces in most geological structures, pure magnesium carbonate is seldom found on its own in larger deposits, a fact that has puzzled geologist for more than a century [7].

In 1908, Neuberg and Rewald tried to synthesise magnesite in alcohol suspensions of MgO [8]. However, it was concluded that $MgCO_3$ cannot be obtained by passing $CO_2$ gas through such suspensions due to the more likely formation of magnesium dimethyl carbonate ($Mg(OCOOCH_3)_2$). Subsequent studies by Kurov in 1961 [9] and Buzágh in 1926 [10] only reiterated the assumption that MgO preferentially forms complex dimethyl carbonates when reacted with $CO_2$ in methanol. A further overview of early works is provided in detail in Text S1and Figure S1. Yet, by changing the synthesis conditions in comparison to what has been described earlier, we here report the successful formation of a magnesium carbonate, hereafter referred to as*Upsalite*, in a reaction between MgO, methanol and $CO_2$ resulting in an anhydrous, micro-mesoporous and large surface area structure. We further show that the moisture sorption of the material is featured by a unique set of properties including an adsorption capacity ~50% larger than that of the hygroscopic zeolite-Y at low relative humidities and with the ability to retain more than 75% of the adsorbed water when the humidity is decreased from 95% to 5% at room temperature. The humid material is easily regenerated to regain its moisture sorption characteristic upon storage at only 95°C.

## Results and Discussion

The synthesis is carried out well below 100°C, while previously reported amorphous structures of magnesium carbonate have been formed at higher temperatures by thermal decomposition of hydrated magnesium carbonates [11]–[14] or of a double salt of magnesium ammonium carbonate [15]. In the current work, $CO_2$ is not bubbled through the methanolic suspension, instead the reaction vessel is pressurised with $CO_2$ to moderate relative pressures (1–3 bar). Initially, the temperature is kept at 50°C in order to facilitate a reaction between MgO and methanol, and after ~ 3 h the temperature is decreased to room temperature. This results in formation of a rigid gel in the reaction vessel after ~ 4 days. When dried in air at 70°C, the gel solidifies and collapses into a white and coarse powder that is primarily X-ray amorphous with traces of unreacted and crystalline MgO, see XRD pattern in Figure 1a. The sharp peaks at 2 θ equal to 43° and 62° originate from the unreacted MgO [16], while the halo peak between 2 θ values of 25° and 40° is indicative of at least one amorphous phase.

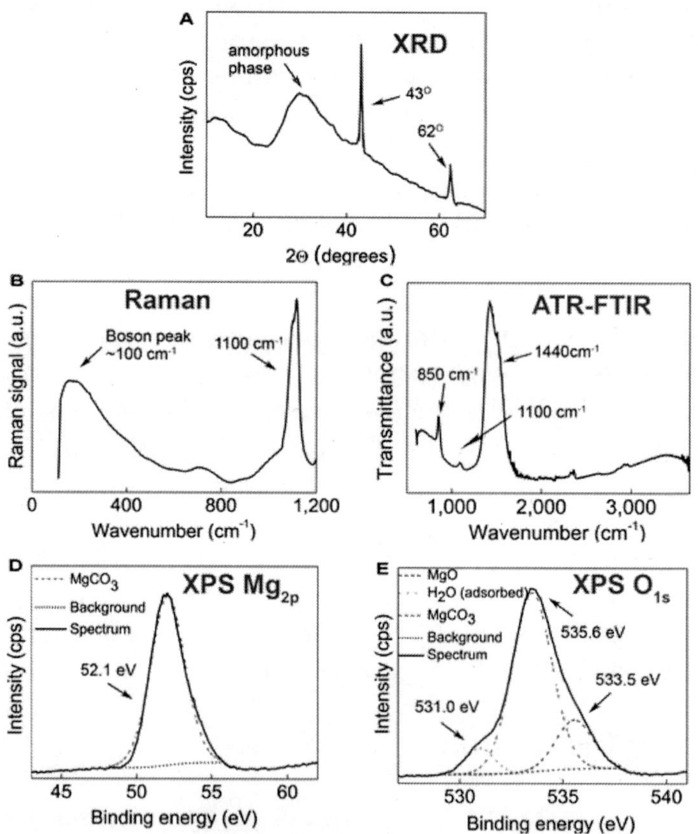


Figure 1. Characterisation of as-synthesised Upsalite and schematic description of the synthesis steps.

**a)** XRD pattern. The halo at 2 θ~30° indicates the presence of at least one amorphous phase and the sharp peaks at higher scattering angles pertain to crystalline MgO. **b)**Raman spectrum. The band observed at ~1100 $cm^{-1}$ stems from vibration of the $CO_3$group and the halo cantered at 100 $cm^{-1}$ is a Boson peak. **c)** FTIR spectrum. The three visible absorption bands (1440 $cm^{-1}$, 1100 $cm^{-1}$ and 850 $cm^{-1}$) are all due to vibrations of the $CO_3$ group. **d)** and **e)** XPS $Mg_{2p}$ and $O_{1s}$ peaks. The $Mg_{2p}$ peak at 52.1 eV and the $O_{1s}$ peak at 533.5 eV stem from $MgCO_3$, the $O_{1s}$ peak at 531.0 eV from MgO and the $O_{1s}$ peak at 535.6 eV from surface adsorbed water. The solid lines represent the measured spectrum. The coloured lines are calculated using the CasaXPS software and represent the fitted curves (obtained using Gaussian-Lorentzian functions) and the subtracted background (obtained using a Shirley function).

doi:10.1371/journal.pone.0068486.g001

Raman spectroscopy reveals that the powder indeed is composed of a carbonate (Figure 1b), where the band at ~1100 $cm^{-1}$ corresponds to vibration of the carbonate group [17]. Moreover, a broad halo, or the so-called Boson peak, with a maximum at ~100 $cm^{-1}$, is further witness to the amorphous character of the powder [18].

When examined with Fourier transform infrared spectroscopy (FTIR, Figure 1c) the material displayed absorption bands at ~1440 $cm^{-1}$, ~1100 $cm^{-1}$ and ~850 $cm^{-1}$, which all correspond to the carbonate group [19]. No water of crystallisation is visible in this spectrum [13], [19]. The anhydrous character of the bulk material is further confirmed by Thermal Gravimetric Analysis (TGA) (see Figure S2).

X-ray photoelectron spectroscopy (XPS) confirms the anhydrous nature of the $MgCO_3$. Energy resolved spectra were recorded for the $Mg_{2p}$ and $O_{1s}$ peaks (Figure 1d,e) which were found to be positioned at 52.1 eV and 533.5 eV, respectively, which is indicative of $MgCO_3$ [20]. Further, the $O_{1s}$ peak does not contain any components for crystal water which, expectedly, would have appeared at 533–533.5 eV [21]. The shoulder seen at 535.6 eV is located between the binding energies for liquid water (539 eV) and ice (533 eV) [22] and is, therefore, representative of surface adsorbed water as previously described for adsorbed water on carbon fibres [23],[24]. The shoulder seen at 531.0 eV shows the presence of MgO in the powder. No presence of $Mg(OH)_2$ was observed in the bulk, which would have resulted in a peak at 532.4 eV [20].

Having proved the formation of amorphous anhydrous $MgCO_3$ by XRD, Raman, FTIR, and XPS, we postulate the following simplified route of synthesis, based on the presence of $HOMgOCH_3$as an intermediate (as confirmed with FTIR in Figure S3) and the necessity of heat treatment in the last synthesis step:

$$MgO + CH_3OH \leftrightarrow HOMgOCH_3 \quad \textbf{(1a)}$$

$$CH_3OH + CO_2 \leftrightarrow CH_3OCOOH \quad \textbf{(1b)}$$

$$CH_3OCOOH + HOMgOCH_3$$

$$\rightarrow H_3COCOOMgOCH_3 + H_2O \quad \textbf{(1c)}$$

$$H_3COCOOMgOCH_3 + H_2O$$

$$\leftrightarrow HOMgOGOOCH_3 + CH_3OH \quad \textbf{(1d)}$$

The presence of a hydroxyl group in the vicinity of the methoxy group in $HOMgOCOOCH_3$makes the compound labile and will therefore favour internal transition to a solvate $MgCO_3 \cdot CH_3OH$ and

$$HOMgOCOOCH_3$$

$$\rightarrow MgCO_3 \cdot CH_3OH$$

loss of methanol upon heating to 70°C:$\rightarrow MgCO_3 + CH_3OH \uparrow (70^{\circ}C)$
**(1e)**

As it is evident from the above cascade scheme, the postulated reaction of $MgCO_3$ formation goes through several steps, some of which are equilibrium reactions, namely 1a and 1b. A schematic description of the reaction steps is presented in Figure 2.

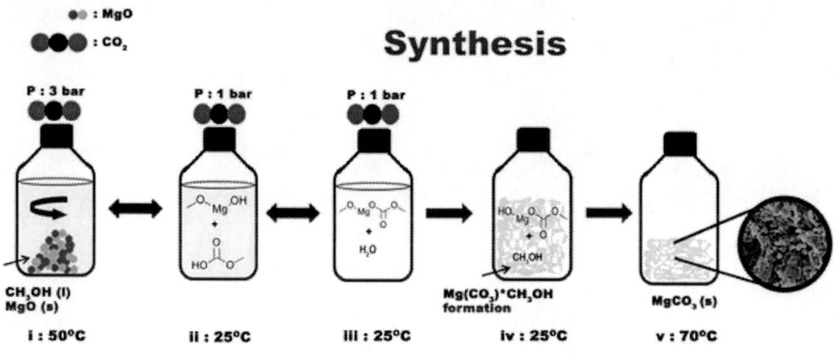


Figure 2. Synthesis of Upsalite.

**i)** In the first step MgO (s) is mixed with methanol under 3 bar $CO_2$ pressure at 50°C. **ii)**After 2.5 h the $HOMgOCH_3$ is formed in the solution, the pressure is lowered to 1 bar and the heating is turned off. At the same time the methanol reacts with the $CO_2$ and forms $CH_3OCOOH$ (methyl hemicarbonic acid). **iii)** $HOMgOCH_3$ reacts with $CH_3OCOOH$ and forms water and $H_3COCOOMgOCH_3$ (methyl esther of magnesium methyl carbonate). At this point the solution changes colour from white to light yellow. **iv)** $H_3COCOOMgOCH_3$ reacts with the water formed in step iii) and forms $HOMgOCOOCH_3$(or $MgCO_3 \cdot CH_3OH$) which upon **v)** heating at 70°C releases $CH_3OH$ and forms $MgCO_3$.

doi:10.1371/journal.pone.0068486.g002

In order to analyse the pore structure and water sorption capacity of Upsalite, $N_2$ and $H_2O$ vapour sorption analyses were performed. Figure 3a shows the $N_2$ sorption isotherm for Upsalite, which exhibits a typical Type 1 shape according to the IUPAC classification [25]. The SSA of 800 $m^2\ g^{-1}$ for the material (Table 1) was derived from such isotherms according to the Brunauer-Emmet-Teller (BET) equation [26], substantially surpassing the SSA of all previously described alkali earth metal carbonates, where crystalline forms of magnesium carbonates typically have a SSA of 4–18 $m^2\ g^{-1}$ [5]. This high SSA places Upsalite in the exclusive class of high surface area nanomaterials including mesoporous silica, zeolites, MOFs, and carbon nanotubes.

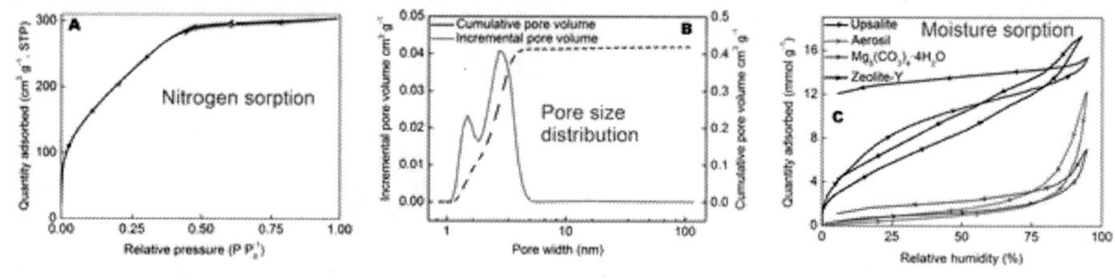


Figure 3. Sorption isotherms and DFT-based pore size distribution for Upsalite.

**a)** $N_2$ sorption isotherm at 77 K. **b)** Incremental pore volume (violet) and cumulative pore volume (blue) obtained from $N_2$ sorption isotherm. **c)** Moisture sorption isotherm at room temperature for Upsalite (blue), $Mg_5(CO_3)_4(OH)_2 \cdot 4H_2O$ (green), Aerosil (red) and Zeolite Y (black). The arrows indicate the direction of the pressure change.

doi:10.1371/journal.pone.0068486.g003

Table 1. Structural and chemical characteristics of Upsalite obtained from $N_2$and $H_2O$ vapour sorption isotherms.

doi:10.1371/journal.pone.0068486.t001

| **Adsorbate** | **$N_2$** | **$H_2O$** |
|---|---|---|
| SSA[a] ($m^2/g$) | 800.0±3.6 | – |
| Total pore volume[b] ($cm^3/g$) | 0.47 | – |
| $w_0$, limiting micropore volume[c] ($cm^3/g$) | 0.280±0.001 | 0.160±0.010 |
| Equivalent surface area in micropores[c] ($m^2/g$) | 549 | 478 |
| Characteristic energy of adsorption[c] (kJ/mol) | 11.4 | 41.0 |
| Modal equivalent pore width[c] (nm) | 1.75 | 1.09 |
| Correlation coefficient of fit[c] | 0.999 | 0.977 |

[a]According to the BET equation

[b]Single point adsorption at $P/P_0 \approx 1$

[c]According to the Dubinin-Astakhov equation

doi:10.1371/journal.pone.0068486.t001

Figure 3c displays the $H_2O$ vapour sorption isotherm for Upsalite and, based on the large amount of $H_2O$ adsorbed at low RHs, it is evident that the material is highly hydrophilic [27]. The limited desorption of moisture from the material when the vapour pressure is reduced from 95% is further proof of the strong interaction between water molecules and the material. It should, however, be noted that no signs of hydrate formation in the material are seen using XRD after the sorption isotherm is completed, and that the sorption isotherm can be repeated with undistinguishable results after heat treatment at moderate temperatures (95°C) under vacuum. This contrasts to the regeneration of moisture sorption properties of, e.g., Zeolites typically requiring heat treatments at temperatures between 150°C and ~600°C.

Further, both the $N_2$ and the $H_2O$ vapour sorption isotherms were analysed in order to establish the microporous properties of the material according to the Dubinin-Astakhov (D-A) model [28], see Table 1. The hydrophilic nature of the material is further reflected in the greater characteristic energy for adsorption of $H_2O$ compared to $N_2$. The discrepancy in the limiting micropore volume ($w_0$) – in which the value obtained from the $N_2$ sorption isotherm is to be regarded as the "true" value – and modal equivalent pore size obtained from the two sorption isotherms is most likely due to site-specific interaction between the $H_2O$ species and the material, not only in the micropores but also on the exterior of the material and in pores larger than 2 nm [29].

Figure 3b shows the incremental and cumulative pore volume obtained through density functional theory (DFT) calculations on the $N_2$ sorption isotherm. From these calculations it infers that about 98% of the pore volume is inherent to pores with a diameter smaller than 6 nm, while the remaining pore volume is made up of pores with a broad size distribution between 8 and 80 nm centred at ~16 nm.. When examined with scanning electron microscopy (SEM), these pores in the larger size range are clearly visible (Figure 4a–b). Furthermore, the highly porous nature of the material is evident from the scanning transmission electron microscopy (STEM) tomography work available in the Supporting Information (see Video S1 and S2 and Figure S4; S4.1). Such three-dimensional reconstructions allow for visualisation of the internal pore structure of the material via a series of slices through the volume. Measurements from these slices, which essentially represent cross-sections, confirm that pore widths are consistently 16 nm, while pore heights vary between 8 and 50 nm. However, these larger pore networks that are visible with SEM and STEM are not noted throughout the entire material, which is consistent with the limited contribution from these pores to the total pore volume as determined by DFT. In the transmission electron microscopy (TEM) image in Figure 4c the smaller pores, dominating the contribution to the total pore volume sensed by nitrogen sorption, can be distinguished. An enlargement of pores was found to take place under the electron beam where the sample was unstable for long periods of time. This enlargement is most likely due to remaining organic groups leaving the sample. A representative image recorded after a longer period (~ 1 min) under the electron beam is shown in Figure S4 (S4.2).

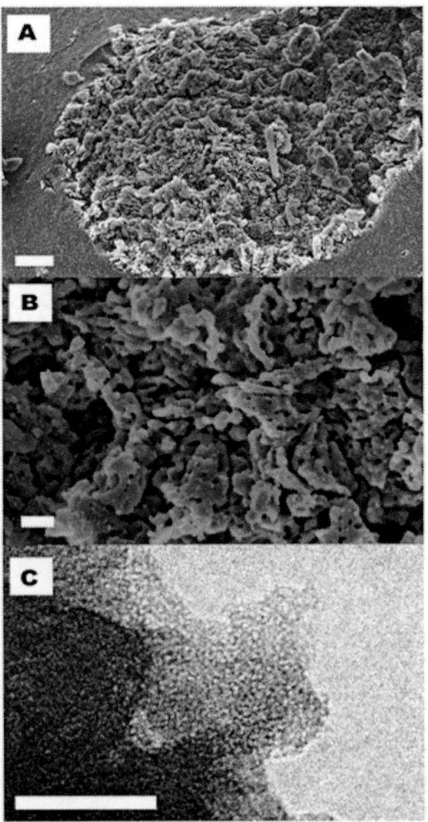


Figure 4. Electron microscopy images of Upsalite.

**a)** SEM micrograph of Upsalite. Scale bar, 1 μm. **b)** Higher magnification SEM of a region in a) clearly showing the textural porosity of the material. Scale bar, 200 nm. **c)**Representative TEM image of Upsalite showing contrast consistent with a porous material. The image is recorded with under-focused conditions to enhance the contrast from the pores. Scale bar, 50 nm.

doi:10.1371/journal.pone.0068486.g004

The water sorption capacity of the material is interesting from an industrial and technological point of view and it is, hencc, compared to three commercially available desiccants, namely fumed silica (SSA: 196 $m^2 g^{-1}$), hydromagnesite (SSA: 38 $m^2 g^{-1}$) and the microporous Zeolite Y (SSA: 600 $m^2 g^{-1}$, silica/alumina ratio 5.2:1), see Figure 3c. For comparison, all samples were degassed at 95°C under vacuum for 10 h prior to analysis. The $H_2O$ vapour adsorption isotherm for Upsalite displays similarities with the hydrophilic zeolite at very low RHs (<1%) and shows an even higher adsorption capacity compared to the zeolite at RHs between 1 and 60%. This behaviour contrasts largely to that of the other two non-porous materials, i.e. fumed silica and hydromagnesite, which mainly adsorb $H_2O$ at RH >60%.

Amorphous magnesium carbonates produced by high temperature thermal decomposition of hydromagnesite, nesquehonite, or magnesium ammonium carbonate double salt have previously been reported to be unstable upon hydration [15], [30]. In particular, instability of the hydromagnesite decomposed material was evident by a weakening of the carbonate bond [30]. Such weakening was observed by a shift towards lower temperatures, as well as by the development of a shoulder and a split into two or more peaks, of the carbonate decomposition peak located above 350°C in differential TGA (dTGA) spectra [30]. In this respect, Upsalite appears to remain stable upon hydration. After 11 weeks of storage at RT and 100% RH no peak split, shoulders or movement of the carbonate decomposition peak towards lower temperatures is observed in dTGA spectra (see Figure S5). In fact, the decomposition peak for the carbonate bond is shifted towards higher temperatures as compared to the as-synthesized material (Figure S2 and S5), indicating a strengthening of the carbonate bond.

## CONCLUSIONS

We report herein the template-free formation of a stable, amorphous magnesium carbonate nanostructure formed at low temperature in a methanol solution of MgO with $CO_2$. The obtained magnesium carbonate is featured with a unique structure of pores almost exclusively in the sub 6 nm size range and extraordinarily high surface area, which has never been reported before, neither for natural nor synthetic magnesium carbonates. The material described in this work is further featured with extraordinary water sorption properties that can be regenerated at temperatures below 100°C. The material is foreseen to find its use in a number of applications including humidity control and delivery systems for therapeutic or volatile agents.

## MATERIALS AND METHODS

### Synthesis

In the current work 4 g magnesium MgO powder was placed in a glass bottle together with 60 ml methanol and a stirring magnet. The solution was put under 3 bar $CO_2$ pressure and heated to 50°C. After approximately 4 hours the mixture was allowed to cool to RT and the carbon dioxide pressure was lowered to 1 bar, and the reaction continued until a gel had formed. When a gel was obtained, the carbon dioxide pressure was removed and the gel was allowed to solidify and dry at ~70°C during 3 days. A schematic description of the synthesis is found inFig. 2.

## CHARACTERISATION

### X-ray diffraction

XRD analysis was performed with a Siemens/Bruker D5000 instrument using Cu-$K_\alpha$ radiation. Samples were ground and put on a silicon sample holder with zero background prior to analysis. The instrument was set to operate at 45 kV and 40 mA.

## RAMAN SPECTROSCOPY

A Reinshaw Ramanscope was used for the Raman studies. The Raman instrument was calibrated with a silicon wafer using the band at 521 $cm^{-1}$ prior to the studies. A 524 nm argon-ion laser with a 10 μm spot size was used for analysis.

### Fourier transform infrared spectroscopy

The FTIR studies were performed on a Bruker IFS 66v/S spectrometer using an Attenuated Total Reflectance (ATR) sample holder from SENSIR. 50 scans were signal-averaged in each spectrum and the resolution was 4 $cm^{-1}$. Before the measurement a background scan was recorded and thereafter subtracted from the spectrum for the sample.

### X-ray photoelectron spectroscopy

The XPS experiments were conducted on a Phi Quantum 200 Scanning ESCA microprobe instrument. Prior to analysis, the samples were sputter cleaned using argon ions for 10 min at 200 V to remove surface adsorbed contaminations. A full spectrum was recorded together with energy resolved spectra for $Mg_{2p}$ and $O_{1s}$. During the acquisition, an electron beam of 20 μA was used together with argon ions to neutralise the non-conducting sample. The peak fittings were made with CasaXPS software, the curves were fitted using Gaussian-Lorentzian functions and the background was subtracted using a Shirley function. The spectra were calibrated against the $O_{1s}$ peak for magnesium oxide (531.0 eV) instead of the $C_{1s}$ peak at 285.0 eV for adventitious carbon, which otherwise is commonly used as a reference. However, in the case of MgO, the binding energy for adventitious carbon is not reliable as reference since hydrocarbons interact with magnesium oxide in a way that shifts the $C_{1s}$ peak randomly making it unsuitable as a reference. Therefore, the $O_{1s}$ peak for MgO (531.0 eV) is instead proposed to be used as an internal reference [31]. The presence of magnesium oxide in the samples was confirmed by XRD analysis prior to the XPS study.

## $N_2$ sorption analysis

Gas sorption measurements were carried out with $N_2$ at 77 K using an ASAP 2020 from Micromeritics. The samples were degassed at 95°C under vacuum for 10 h prior to analysis. The SSA was determined by applying the BET equation [26] to the relative pressure range of 0.05–0.30 for the adsorption branch of the isotherm. The D-A equation was employed on the appropriate pressure region for adsorption in micropores. The BET and D-A calculations were performed with the ASAP 2020 V3.04 software from Micromeritics delivered together with the analysis equipment. The pore size distribution was determined using *DFT analysis* carried out with the DFT Plus software from Micrometrics using the model for $N_2$ at 77K for slit-shape geometry with low regularisation ($\lambda$ = 0.005). The standard deviation of the DFT fit was 2.037 $cm^3/g$.

## Scanning electron microscopy

For the SEM analyses, a Leo 1550 instrument from Zeiss equipped with an in-lens detector was used. Prior to the studies, the samples were cooled with liquid $N_2$, crushed and put on a stub holder with double-sided carbon tape. As a last step prior to analysis the sample was sputter coated with a thin layer of gold/palladium.

## Thermal gravimetric analysis

TGA analysis was carried out under a flow of air in an inert alumina cup with sample sizes of approximately 15 mg. The samples were heated from RT to 700°C with a heating ramp of 10°C $min^{-1}$ using a Thermogravimetric analyser from Mettler Toledo, model TGA/SDTA851e.

## Water vapour sorption

An ASAP 2020 instrument from Micromeritics was used for the water sorption studies. Prior to analysis the samples were degassed at 95°C under vacuum for 10 h. The D-A equation was employed on the appropriate pressure region for adsorption in micropores. The affinity coefficient ($\beta$) for water was set to 0.2 in the D-A calculations, which has been shown to be an appropriate value for analysis of polar surfaces [32].

## Scanning transmission electron microscopy and electron tomography

Scanning transmission electron microscopy (STEM) samples were prepared by dispersing the powder in ethanol and placing 20 μl on a Quantifoil® TEM grid. Experiments were performed on an FEI Tecnai F20 (FEI Company, The Netherlands) operated at 200 kV. Images were recorded on a high-angle annular dark-field detector (HAADF). The Dual-Axis Tomography Holder Model 2040 (Fischione Instruments, PA, USA) was used in a linear tilt scheme to acquire a single-axis tilt-series with image acquisition increments of 2°. Automated focusing, image shifting, and acquisition of HAADF STEM images over an angular range of ±62° were achieved using the Explore3D software (FEI Company, The Netherlands). The 3D reconstructions were computed using a simultaneous iterative reconstruction technique, with 20 iterations, in Inspect3D (FEI Company, The Netherlands). Models for 3D visualisation were created in Amira Resolve RT FEI (Visage Imaging Inc., USA).

## Transmission Electron Microscopy (TEM)

HRTEM images were taken with a JEOL-3010 microscope, operating at 300 kV (Cs 0.6 mm, resolution 1.7 Å). Images were recorded using a CCD camera (model Keen View, SIS analysis, size 1024×1024, pixel size 23.5×23.5 μm) at 30 000–100 000× magnification using low-dose conditions on as-crushed samples.

## Acknowledgments

We thank Prof. Peter Lazor for assistance with interpretation of the Raman spectra, Dr. Alfonso E. Garcia-Bennett for recording the TEM images and Dr. Martin Sjödin for initial discussions regarding the reaction mechanism.

## AUTHOR CONTRIBUTIONS

Conceived and designed the experiments: JF AM MS. Performed the experiments: SF KG JF. Analyzed the data: JF AM SF MS KG. Contributed reagents/materials/analysis tools: JF AM SF MS KG. Wrote the paper: JF AM SF MS KG.

## REFERENCES


1. Fadeel B, Kasemo B, Malmsten M, Strømme M (2010) Nanomedicine: Reshaping clinical practice. J. Intern. Med. 267: 2–8. doi: 10.1111/j.1365-2796.2009.02186.x
2. European Commission (2012) A European strategy for Key Enabling Technologies – A bridge to growth and jobs. Brussels: COM 341 final.
3. Mihranyan A, Ferraz N, Strømme M (2012) Current status and future prospects of nanotechnology in cosmetics. Prog. Mater. Sci. 57: 875–910. doi: 10.1016/j.pmatsci.2011.10.001
4. Varshney HM, Shailender M (2012) Nanotechnology; Current Status In Pharmaceutical Science: A Review. Int. J. Therap. Appl. 6: 14–24. doi: 10.1016/j.pmatsci.2011.10.001
5. Truitt B (2009) Magnesium Carbonate. In Handbook of Pharmaceutical Excipients, 6th ed.; Rowe R, Sheskey P, Quinn M, Eds.; Pharmaceutical Press: London; 397–400.
6. Deelman JC (2011) In Low Temperature Formation of Dolomite and Magnesite [Online] version 2.3 ed.; Compact Disc Publications: Eindhoven, 2011. Available:www.jcdeelman.demon.nl. Accessed 2012 Apr 18.
7. Pohl W (1990) Genesis of Magnesite Deposits - Models and Trends. Int. J. Earth. Sci. 79: 291–299. doi: 10.1007/bf01830626
8. Neuberg C, Rewald B (1908) Ueber Kolloide Und Gelatinöse Verbindungen Der Erdalkalien. Colloid Polymer Sci. 2: 354–357. doi: 10.1007/bf01830626
9. Kurov VI (1961) Alkyl Carbonate Salts.5. Methyl Carbonate Salts of Bivalent Metals. Russ. J. Gen. Chem. 31: 9–11. doi: 10.1007/bf01830626
10. Buzágh A (1926) Ueber Kolloide Lösungen Der Erdalkalikarbonate. Kolloid Z. 38: 222–226. doi: 10.1007/bf01830626
11. Khan N, Dollimore D, Alexander K, Wilburn FW (2001) The Origin of the Exothermic Peak in the Thermal Decomposition of Basic Magnesium Carbonate. Thermochim. Acta 367–368: 321–333. doi:

10.1007/bf01830626

12. Hashimoto H, Tomizawa T, Mitomo M (1968) Exothermic Process in the Differential Thermal Analysis Curves of Basic Magnesium Carbonate. Kogyo Kagaku Zasshi 71: 480–484. doi: 10.1246/nikkashi1898.71.4_480
13. Sawada Y, Uematsu K, Mizutani N, Kato M (1978) Thermal Decomposition of Hydromagnesite 4 $MgCO_3$•$Mg(OH)_2$•4 $H_2O$ under Different Partial Pressures of Carbon-Dioxide. Thermochim. Acta 27: 45–59. doi: 10.1246/nikkashi1898.71.4_480
14. Choudhary VR, Pataskar SG, Gunjikar VG, Zope GB (1994) Influence of Preparation Conditions of Basic Magnesium Carbonate on Its Thermal-Analysis. Thermochim. Acta 232: 95–110. doi: 10.1246/nikkashi1898.71.4_480
15. Dell RM, Weller SW (1959) The Thermal Decomposition of Nesquehonite $MgCO_3$•3 $H_2O$ and Magnesium Ammonium Carbonate $MgCO_3$•$(NH_4)2CO_3$•4 $H_2O$. Trans Faraday Soc 55: 2203–2220. doi: 10.1246/nikkashi1898.71.4_480
16. Clark CB (1946) X-Ray Diffraction Data for Compounds in the System CaO-MgO-$SiO_2$. J. Am. Ceram. Soc. 29: 25–30. doi: 10.1246/nikkashi1898.71.4_480
17. Rutt HN, Nicola JH (1974) Raman Spectra of Carbonates of Calcite Structure. J. Phys. C Solid State Phys. 7: 4522–4528. doi: 10.1088/0022-3719/7/24/015
18. Gouadec G, Colomban P (2007) Raman Spectroscopy of Nanomaterials: How Spectra Relate to Disorder, Particle Size and Mechanical Properties. Prog. Cryst. Growth Charact. Mater. 53: 1–56. doi: 10.1088/0022-3719/7/24/015
19. Raade G (1970) Dypingite, a New Hydrous Basic Carbonate of Magnesium, from Norway. Am. Mineral. 55: 1457–1465. doi: 10.1088/0022-3719/7/24/015
20. Santamaria M, Di Quarto F, Zanna S, Marcus P (2007) Initial Surface Film on Magnesium Metal: A Characterization by X-Ray Photoelectron Spectroscopy (XPS) and Photocurrent Spectroscopy (Pcs). Electrochim. Acta 53: 1314–1324. doi: 10.1088/0022-3719/7/24/015
21. Splinter SJ, Mcintyre NS, Lennard WN, Griffiths K, Palumbo G (1993) An Aes and Xps Study of the Initial Oxidation of Polycrystalline Magnesium with Water-Vapor at Room-Temperature. Surf. Sci. 292: 130–144. doi: 10.1088/0022-3719/7/24/015
22. Siegbahn K, Nordling C, Johansson G, Hedman J, Hedén PF, et al..

(1969) Esca Applied to Free Molecules. North-Holland Publishing Company: Amsterdam.

23. Xie YM, Sherwood PMA (1990) X-Ray Photoelectron Spectroscopic Studies of Carbon-Fiber Surfaces.11. Differences in the Surface-Chemistry and Bulk Structure of Different Carbon-Fibers Based on Poly(Acrylonitrile) and Pitch and Comparison with Various Graphite Samples. Chem. Mater. 2, 293–299.
24. Gardner SD, Singamsetty CSK, Booth GL, He GR, Pittman CU (1995) Surface Characterization of Carbon-Fibers Using Angle-Resolved Xps and Iss. Carbon 33: 587–595. doi: 10.1016/0008-6223(94)00144-o
25. Sing KSW, Everett DH, Haul RAW, Moscou L, Pierotti RA, et al. (1985) Reporting Physisorption Data for Gas Solid Systems with Special Reference to the Determination of Surface-Area and Porosity (Recommendations 1984). Pure Appl. Chem. 57: 603–619. doi: 10.1016/0008-6223(94)00144-o
26. Brunauer S, Emmett PH, Teller E (1938) Adsorption of Gases in Multimolecular Layers. JACS 60: 309–319. doi: 10.1016/0008-6223(94)00144-o
27. Pires J, Pinto ML, Carvalho A, de Carvalho MB (2003) Assessment of Hydrophobic-Hydrophilic Properties of Microporous Materials from Water Adsorption Isotherms. Adsorption 9: 303–309. doi: 10.1023/a:1026219813234
28. Dubinin MM, Astakhov VA (1971) Description of Adsorption Equilibria of Vapors on Zeolites over Wide Ranges of Temperature and Pressure. In Molecular Sieve Zeolites-Ii, American Chemical Society: Washington, D C. 102: 69–85. doi: 10.1023/a:1026219813234
29. Verhoeven L, Lodewyckx P (2001) In Comparison of Dubinin-Radushkevich Micropore Volumes Obtained from $N_2$, $CO_2$ and $H_2O$-Adsorption Isotherms, Carbon 2001 proceedings, Lexington (KY, USA), American Carbon Society: Conference proceeding avalibe from: http://acs.omnibooksonline.com/data/papers/2001_10.2.pdf.
30. Botha A, Strydom CA (2003) Dta and FT-IR Analysis of the Rehydration of Basic Magnesium Carbonate. J. Therm. Anal. Calorim. 71: 987–995. doi: 10.1023/a:1026219813234
31. Ardizzone S, Bianchi CL, Fadoni M, Vercelli B (1997) Magnesium Salts and Oxide: An Xps Overview. Appl. Surf. Sci. 119: 253–259. doi: 10.1023/a:1026219813234
32. Dastgheib SA, Karanfil T (2005) The Effect of the Physical and Chemical Characteristics of Activated Carbons on the Adsorption Energy and Affinity Coefficient of Dubinin Equation. J. Colloid Interface Sci. 292: 312–321.

# Chapter 11

# INFLUENCE OF CUTICLE NANOSTRUCTURING ON THE WETTING BEHAVIOUR/ STATES ON CICADA WINGS

Mingxia Sun[1], Aiping Liang[1] *, Gregory S. Watson[2], Jolanta A. Watson[2], Yongmei Zheng[3]*, Jie Ju[4], Lei Jiang[4]

[1]Key Laboratory of the Zoological Systematics and Evolution, Institute of Zoology, Chinese Academy of Sciences, Beijing, China

[2]School of Pharmacy and Molecular Sciences, James Cook University, Townsville, Queensland, Australia

[3]Key Laboratory of Bio-Inspired Smart Interfacial Science and Technology of Ministry of Education, School of Chemistry and Environment, Beihang University, Beijing, China, [4]Center of Molecular Sciences, Institute of Chemistry, Chinese Academy of Sciences, Beijing, China

## ABSTRACT

The nanoscale protrusions of different morphologies on wing surfaces of four cicada species were examined under an environmental scanning electron microscope (ESEM). The water contact angles (CAs) of the wing surfaces were measured along with droplet adhesion values using a high-sensitivity microelectromechanical balance system. The water CA and adhesive force measurements obtained were found to relate to the nanostructuring differences of the four species. The adhesive forces in combination with the Cassie-Baxter and Wenzel approximations were used to predict wetting states of the insect wing cuticles. The more disordered and inhomogeneous surface of the species *Leptopsalta bifuscata* demonstrated a Wenzel type wetting state or an intermediate state of spreading and imbibition with a CA of 81.3° and high adhesive force of 149.5 μN. Three other species (*Cryptotympana atrata, Meimuna opalifer* and *Aola bindusara*) exhibited nanostructuring of the form of conically shaped protrusions, which were spherically capped. These surfaces presented a range of high adhesional values; however, the CAs were highly hydrophobic (*C. atrata* and *A. bindusara*) and in some cases close to superhydrophobic (*M. opalifer*). The wetting states of *A. bindusara, C. atrata* and *M. opalifer* (based on adhesion and CAs) are most likely represented by the transitional region between the Cassie-Baxter and Wenzel approximations to varying degrees.


## INTRODUCTION

Research on the superhydrophobic and self-cleaning properties of natural materials, such as the famous lotus leaf, has been undertaken for more than ten years [1], [2]. The original impetus for studies in this area stems from the low adhesion observed between lotus leaves and water or other contaminants [1], [3]. It has been shown, however, that the wetting properties on different regions of the leaf vary. For instance the flat folds around the margin of the lotus leaf have a much larger contact angle (CA) hysteresis (CAH) than that of the upper surface of the lotus leaf including the micro-papillae [4]. There are a number of other natural superhydrophobic surfaces which also demosntsrate a large CAH. The rose petal surface, for example, consists of hierarchical micropapillae and nanofolds, which provide a sufficient roughness for superhydrophobicity and yet at the same time a high adhesive force with water[5]. In this case the pitch values of microstructures and density of nanostructures play an important role [6]. For scallions and garlic, hydrophobic defects result in contact line pinning and high CAH [7].

In general, a range of factors influence the adhesive properties of solid surfaces such as chemical compositions, density [8], real area of contact, surface energy effects [9], surface roughness [10] and the apex geometry [11]. A contributing aspect for the resulting high adhesion between some surfaces and water (besides the capillary force [12] and the negative pressure produced by the volumes of sealed air [11]) has been ascribed to van der Waals forces [11]–[13].

For some biological samples, the heterogeneous nature of the surface also plays an important role in adhesive properties [6]. Many naturally occurring nano-structures have demonstrated functional efficiencies which are superior to man-made technologies. One of the most noteworthy nano-composite materials is the insect cuticle [14]. Recently micro- and nano-structures found on insect cuticle have been shown to exhibit a range of impressive and remarkable properties such as superhydrophobicity, directed wetting, self cleaning and ultra-low adhesion [15]–[26]. The cuticle on the wings of insects demonstrates a wide variety of small scale structuring. The cicada wing is a prime example. A range of interesting properties have been demonstrated on the surfaces of cicada wings with functions and functional efficiencies related to the structure parameters (shape, size, spacing etc). Current studies have focused on a number of aspects such as wettability, antireflection, self cleaning, particle adhesion, antimicrobical activity, cell growth platforms,, material properties and biomimetic fabrication of the nanostructures [27]–[38]. Indeed some of these atributes (self-cleaning, antimicrobical, cell growth, antirelfection) suggest that biomimetic fabrication may potentially have a wide variety of applications ranging from clinical biomaterials (e.g., ocular tissue engineering strategies), antibacterial surfaces/implants and self-cleaning medical based surfaces [37], [38]. We have recently reported on the interaction of water with cicada forewings[39]. We showed significantly different wettabilities associated with distinct differences in surface patterning of nanostructures. The primary focus of this study is an investigation of the wetting and adhesive properties of the wing membranes of four specific cicada species.The study examines the adhesional dependence with the different membrane morphologies and analyzes the mechanism of high adhesion through modeling to ascertain the most likely wetting states.

**Results and Discussion**

## Surface morphology and wettability

Unlike the hierarchical micro- and nanostructuring of the lotus leaf and rose petal, the cicada wing surfaces exhibited a single level of roughness where small nanoscale structures were observed. Fig. 1 shows the fine protrusions and different morphologies found on the four species of cicada wing surfaces studied. The species *Leptopsalta bifuscata* has nanostructuring comprising of a dome shape resulting in spherically capped surface structures (Fig. 1A). Some of the taller structures exhibit a cylindrical shape which is spherically capped. This type of surface represents the most disordered/inhomegeous of the structuring in this study. The relevant structure parameters are: average basal diameter (*d*) of 90 nm, basal spacing (*s*) of 117 nm and height (*h*) of 200 nm. On this type of surface the wettability is exhibited as hydrophilic with a low CA of 81.3° (inset in Fig. 1Aii).

The other three cicada species demonstrate a more homogeneous surface topography with conical shaped nanostructures and display a stronger hydrophobicity than the cylindrically shaped structuring of *L. bifuscata* (Fig. 1, Table 1). The cicada species *Aola bindusara* (Fig. 1B) clearly shows a more ordered structuring than *L. bifuscata* with a more consistent structure height (lower SD) and a higher CA of 135.5°. The structuring, although more conical in shape, could still be approximated as spherical at the apex of structures. The spherical apex appears swollen on most of the nano-structures. The other two cicada species (*Meimuna opalifer* with a CA of 143.8° and *Cryptotympana atrata* with CAs of 132.7° or 137.9°) exhibit more conically shaped protrusions as seen in Figs. 1C and 1D, respectively. These cicadas also demonstrate a much higher structure height (*h*) on the wing surface compared to *A. bindusara* (up to twice the height, see Fig. 1 and Table 1). Structures on *M. opalifer* exhibit significantly larger *d* values (ca. 148 nm) and smaller *s* values (ca. 48 nm) when compared to the other species under investigation. So, the slightly swollen apex and the significantly larger basal diameter and height coupled with the small spacing of protrusions appears to efficiently increase hydrophobicity of the regularly patterned cicada surfaces. The swollen apex shape may be an attempt to reduce weight and material.

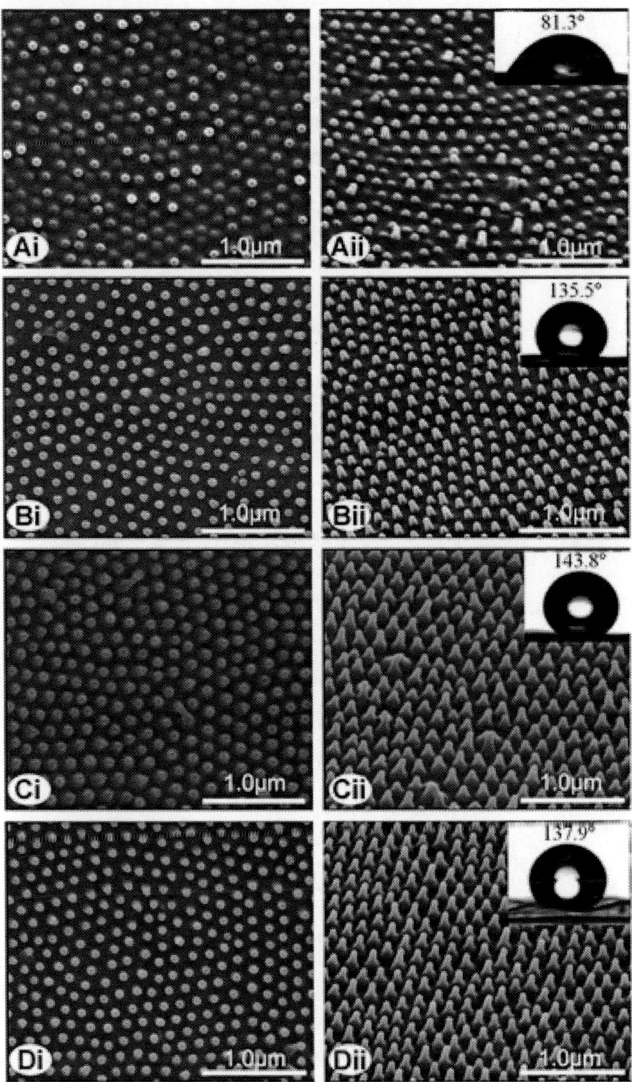


Figure 1. SEM images of four species of cicada wing surfaces.

(A) *Leptopsalta bifuscata*; (B) *Aola bindusara*; (C) *Meimuna opalifer*; (D) *Cryptotympana atrata*. The insets are the optical images of water droplets on the wing surfaces. Subsections (i) show a top view and (ii) show a 30° inclined view of the wing membranes.

doi:10.1371/journal.pone.0035056.g001

Table 1. Mean values and standard deviations (in brackets) of nanostructure parameters including diameter (d), spacing (s), height (h) and the density of protrusions, the adhesive force, measured contact angles (CAs) and calculated contact angle hysteresis (CAH), θw and θc based on the Wenzel and Cassie-Baxter wetting states on dry wing surfaces of four species of cicadas collected on different dates.

doi:10.1371/journal.pone.0035056.t001

| **Species** | **Collection dates** | ***d*** | ***s*** | ***h*** | **Density** | **Force** | **Measured** | **Calculated** | | |
|---|---|---|---|---|---|---|---|---|---|---|
| | **D.M.Y** | **nm** | **nm** | **nm** | **1/10000 μm²** | **μN** | **CAs°** | **CAH°** | $\theta_w$° | $\theta_c$° |
| *L. bifuscata* | 22.8.1964 | 90(5) | 117(13) | 200(52) | 30 | 149.5 | 81.3(8.3) | 9.9 | 133.9 | 149.3 |
| *A. bindusara* | 29.6.1956 | 84(4) | 91(13) | 234(18) | 42 | 123.0 | 135.5(5.2) | 1.0 | 157.4 | 146.0 |
| *M. opalifer* | 23.7.1998 | 148(6) | 48(5) | 418(38) | 33 | 131.3 | 143.8(6.0) | 3.3 | - | 125.2 |
| *C. atrata* | 25.8.2010 | 85(5) | 90(8) | 462(34) | 42 | 170.0 | 137.9(1.9) | 1.5 | - | 145.6 |
| | 11.8.1951 | 95(5) | 90(8) | 410(49) | 37 | 171.9 | 32.7(4.0) | 2.5 | - | 143.5 |

**Footnote:** -Values fall outside the boundaries of the Wenzel Equation. Height values were calculated from inclination of the surface by 30°.
doi:10.1371/journal.pone.0035056.t001

Protrusion height differences were also observed between *C. atrata* samples collected in 2010 and 1951. The dry sample collected in 2010 showed a height of 462 nm compared to the dry sample (ca. 410 nm) collected in 1951 (Table 1). This difference may be due to the hydration/age effects (weathering and/or removal of the outermost layering). The differences in the structural parameters for samples of this age can also be representative of variation within the species or indicate a sub-species.

When compared to the other species (including the older, dry samples of the same species), the fresh wing surface of the cicada *C. atrata* showed the strongest hydrophobicity with a CA value of 147.4±6.4° (Fig. S1) approaching superhydrophobic values. In contrast, the naturally dried samples had a measured CA of ca. 137.9° (Table 1). This phenomenon is possibly due to changes in the chemical constituents existing on the wing surfaces. The wing surface membrane consists of a wax layer [39], and this wax layer may degrade over time yielding a more hydrophilic chemistry. Alternatively, the decreased structure height may also contribute to the lower CA, although this effect is expected to contribute less as seen by comparison of the CAs on dry samples of *M. opalifer* and *C. atrata* (both these samples have similar nano structure dimensions). These observed differences (fresh and dried samples) will be explored in future studies.

## Surface CA hysteresis and adhesive force

The wing surfaces of the four species of cicadas studied showed different wettabilities. Water droplets of 3 μL stayed pinned to all wing surfaces, including fresh and dry samples, even when they were inverted (Fig. S2). This demonstrates that the weight of the water droplet is small in comparison to the surface adhesional force. When the volume of the droplet was increased to 10 μL, the droplet on the surface of *A. bindusara* fell when tilted to ca. 65°, with the weight sufficient to overcome a weak pinning of the contact line on the corner of the protrusions. Interestingly, the CAH measurements were varied on the cicada samples (see Table 1) with the highest value recorded on *L. bifuscata* suggesting that the surface may exhibit one of the highest adhesive forces. The maximum volume of a water droplet which remained hanging was 42 μL on the dry sample of *C. atrata* collected in the year 2010 (Fig. 2). This highlights that the extent of the adhesive force between water and the cicada wing surface can be quite large.

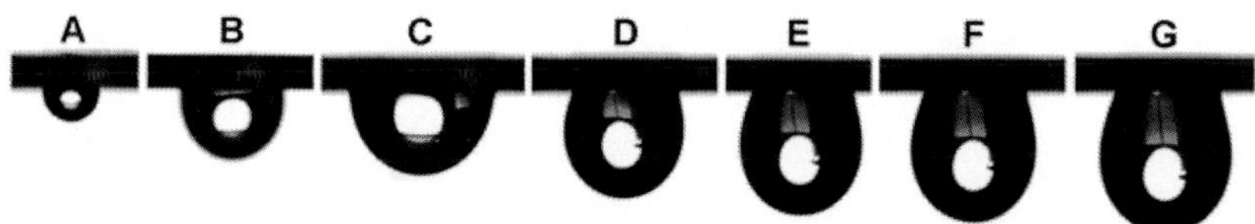


Figure 2. Optical images of different volumes of water droplets hanging on the dry wing surface of cicada Cryptotympana atrata collected in the year 2010.

(A) 3 μL; (B) 20 μL; (C) 30 μL; (D) 35 μL; (E) 37 μL; (F) 39 μL; (G) 42 μL.

doi:10.1371/journal.pone.0035056.g002

To determine the magnitude and differences of adhesion among these cicada wing surfaces the adhesive force was measured using a high-sensitivity microelectromechanical balance system. The method utilized in this study to assess the adhesive force between a solid surface and water can be also used to analyse a number of other solid/liquid interfacial interactions [10]. As shown in Fig. 3, the fresh sample of *C. atrata* resulted in the lowest adhesive force of only 106.3 μN. This is in contrast to the dried samples which exhibited the largest adhesive forces recorded, ca. 170.0 μN and 171.9 μN for the samples collected in 2010 and 1951, respectively (Table 1). The dried samples of *C. atrata* also displayed similar wetting properties with CAs of 137.9° and 132.7°, respectively, despite the different collection dates. Thus changes in wetting properties appear to occur rapidly in the first 3 months, with no significant differences beyond this. The CAH was suprisingly low on the *C atrata* membranes considering the larger magnitude of the adhesion forces in comparision to the other species. Thus a more anisotropic force relationship in relation to water droplet movement appears to be present on this cicada surface. This suggests that the contact area of water with the surface is large enough to promote high adhesion whilst allowing lateral movement to be less hindered. This may potentially be due to the high structure density combined with the different structure morphology (high conically shaped protrusions yeilding large contact area/large adhesion with water droplets) in comparision to the other ciacda structures. The gentle tapering of the structures may be one of the contributing factors which promotes low CAH values.

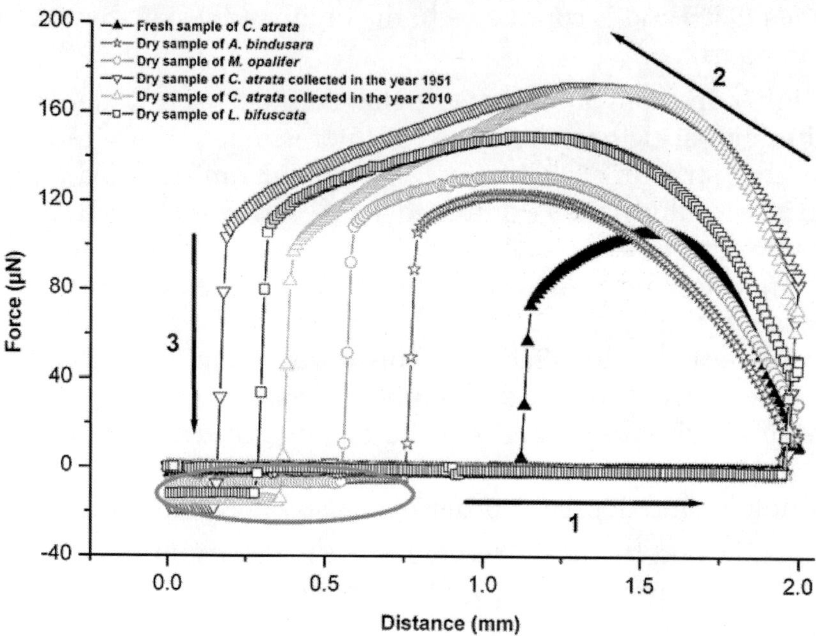


Figure 3. Force-distance curves recorded before and after the water droplet makes contact with the cicada wing.

Process 1: the cicada wing surface approaches the water droplet. Process 2: the cicada wing surface leaves the water droplet after contact. Process 3: the cicada wing surface breaks away from the water droplet. The negative values in the circled section indicate an amount of water still remaining on the wing surfaces when separating from each other.

doi:10.1371/journal.pone.0035056.g003

Among all the dry samples, the surface of *A. bindusara* (Fig. 1B) showed a minimum adhesion value of the order of 123.0 μN (Table 1). It also showed the lowest CAH of all the samples. These lower values can explain why a water droplet of 10 μL cannot remain pinned to the surface when the wing membrane is inverted. The hydrophilic surface of *L. bifuscata* (CA of 81.3°) (Fig. 1A) yielded an adhesive force of 149.5 μN. This value is greater than that of *M. opalifer* (131.3 μN) which is the most hydrophobic dry surface with a CA of 143.8°. The CAH value measured on *M. opalifer* was lower (3.3°) than that of *L.*

*bifuscata* (9.9°) and is conducive to the higher adhesion measured on the later surface.

High adhesion on the rose petal, has been attributed to low height and structure density [6]. Interestingly, the cicada with the higher structure height (*C. atrata* ca. 462 nm) showed a higher adhesional value when compared to *A. bindusara* which also has the same structure density (42×1/10000 $\mu m^2$). However, the surface of *L. bifuscata* which had the lowest protusion density (30×1/10000 $\mu m^2$) but a similar protusion height to *A. bindusara*, resulted in a larger adhesion value (Table 1). This observation is most probably due to the more disordered and inhomogeneous landscape of the *L. bifuscata* cuticle. It appears that the height and density are not the only determining factors for the adhesional contact. The overall morphology and degree of order may also play an important role.

As indicated by the circled region in Fig. 3, as the water droplet breaks away during process 3, the negative values of adhesive forces on the vertical axis indicate that a small amount of water remains pinned on the wing during the retraction cycle. The larger the negative value, the greater the amount of water which remained adhered to the wing. The smallest adhesion value (106.3 μN) on the fresh sample of *C. atrata*, indicated that very little water remained (no negative retract forces in Fig. 3). The dry sample of *C. atrata* collected in the year 1951 however, revealed the greatest amount of water was retained on the wing membrane (i.e., the largest negative retract force value). This surface also displayed the largest measured adhesion of 171.9 μN.

## Wetting states based on water interactions

From the viewpoint of force and energy dissipation, when contacting with a solid, the surface energy of a drop is gradually lowered until ultimately reaching an equilibrium position [40]. On contacting the protrusions, water droplets tend to lower the center of the mass of the liquid, which surface tension opposes. At this moment, the downward gravitational force of the water droplet exceeding the surface tension allows it to move along the wall of protrusions. Different contact areas of water and protrusions result in different interaction forces. The larger the force exerted on the protrusions, the greater the accumulative energy consumed by the reaction force produced by the protrusions. Consequently, if it becomes more diffcult for water to sink further into the structuring, the contact angle will increase.

A number of theories purport to describe the effect of surface roughness on hydrophobicity. As many as six wetting states among two different substances, such as solid-liquid and/or biological surfaces, have been reported [5], [41]. In order to attempt to understand the wetting states of the cicada wing surfaces and the different adhesive forces, we have calculated the predicted contact angles using the Wenzel and Cassie-Baxter models. These two regimes represent two extreme states. The Wenzel model [42] makes the assumption that, when a liquid drop is placed on a surface consisting of protrusions, the liquid will fill the open spaces, as shown in Fig. 4A. This model predicts that roughness of the surface reinforces both hydrophobicity and hydrophilicity. Cassie and Baxter [43] on the other hand, consider the microstructures to be a heterogeneous surface composed of solid and air (Fig. 4B). The crucial assumption is that the space between asperities will remain filled with air with the droplet sitting at the very top of the surface features.

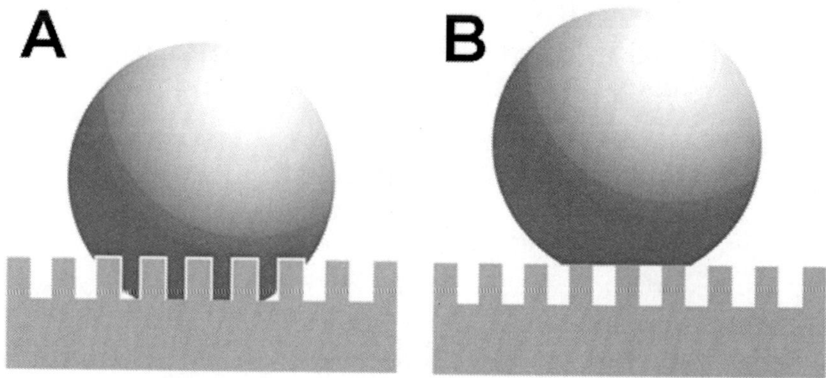


Figure 4. Two wetting states of a liquid on the rough solid surface.

(A) Wenzel model; (B) Cassie-Baxter model.

doi:10.1371/journal.pone.0035056.g004

In the Wenzel state (impregnating wetting regime) the CAH is large, while in the Cassie-Baxter state CAH will be low. Low CAH

also implies low adhesion [6]. The Wenzel and Cassie-Baxter models describe static droplets at equilibrium and allow calculation of the contact angle for the two conditions [42], [43], $\cos\theta_w = \gamma\cos\theta_0$ **(1)**
$\cos\theta_c = \varphi(1+\cos\theta_0) - 1$ **(2)**

where $\gamma$ is the roughness factor (the ratio of actual area to geometry projected area of surface), $\varphi$ is solid fraction in contact with the liquid, $\theta_0$ is the contact angle on a smooth surface of the same material, $\theta_w$ and $\theta_c$ are the apparent contact angles on a rough surface. Thus, the roughness factor ($\gamma$) of wing surfaces was calculated using the following equation (3):

$$\gamma = \frac{(d+s)^2 + 4dh}{(d+s)^2} \quad \textbf{(3)}$$

and the fraction ($\varphi$) was determined using equation (4):

$$\varphi = \frac{d^2}{(d+s)^2} \quad \textbf{(4)}$$

where $d$, $s$ and $h$ are the diameter, spacing and height of protrusions, respectively. Given $\theta_0$ = 105° [15], and substituting $\gamma$ and $\varphi$ into the two model equations, the values of $\cos\theta_w$ and $\cos\theta_c$ were obtained (Table S1). In Table 1, the density of protrusions is defined as the average number of protrusions in the area of 100×100 $\mu m^2$.

Of the cicada wing surfaces studied here, *L. Bifuscata* demonstrated the lowest hydrophobicity with a measured CA of 81.3°. This value is much lower than the one predicted by both the Wenzel (133.9°) and Cassie-Baxter states (149.3°) (Table 1). The contact conditions (based on high adhesion and calculated CA) are more closely aligned with the Wenzel approximation (Fig. 5A), however this may only reflect the wetting state directly beneath the water droplet. The measured low CA may also be a consequence of a thin water film developing in the cuticle nano-texture where the drop sits upon a mixture of solid and liquid. The penetration front spreading beyond the drop is most likely a consequence of the disordered and inhomogeneous surface patterning of the cuticle. Thus the patterning may contribute to this equilibrium wetting state, i.e., the intermediate state of spreading and imbibition due to hemi-wicking [44].

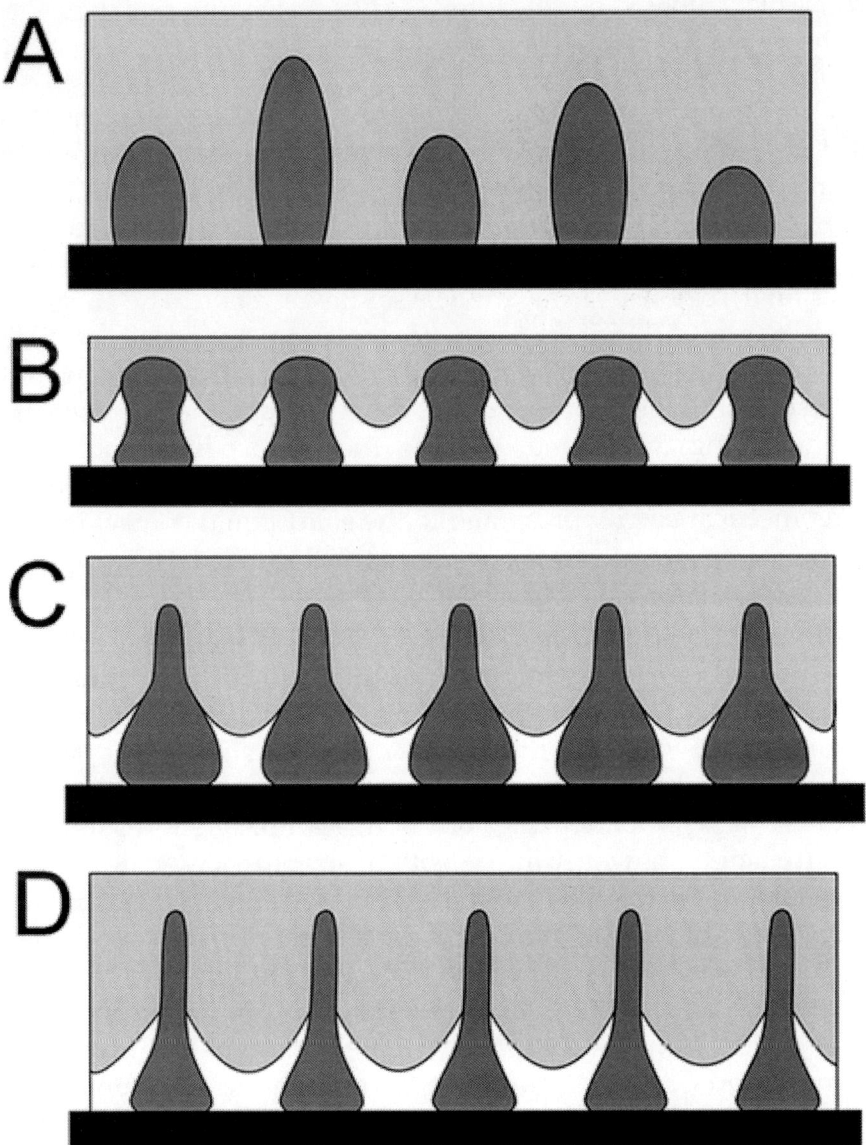


Figure 5. Schematic illustrations of a droplet of water in contact with the cicada wing surface.

(A) *Leptopsalta bifuscata*; (B) *Aola bindusara*; (C) *Meimuna opalifer*; (D) *Cryptotympana atrata*. The dark grey areas represent protrusions, the light grey areas show the water droplet remaining on the surfaces and the black lines show the interface of solid, liquid and gas. doi:10.1371/journal.pone.0035056.g005

A transitional state between Wenzel and Cassie-Baxter states could be expected on the other species from the measured and predicted CAs and adhesional values (see Table 1). The high contact angle oberved on *A. bindusara* combined with the lower adhesion value suggests that the surface structuring is more likely to be in the transitional region nearing the Cassie-Baxter state. The measured contact angle of 135.5° is closer to the predicted value of 146.0° from the Cassie-Baxter model (Table 1). The slightly swollen tops of protrusions may aid in preventing the water droplets from contacting the underlying membrane, keeping the contact line high on the protrusions (Fig. 5B). Thus an energy hurdle is required to overcome the geometrical barrier presented to the solid-liquid contact line.

The larger diameter, smaller spacing (at the base) and greater height of protrusions on the surface of *M. opalifer*, achieved a near superhydrophobic surface. The high CA and moderate adhesion value indicates surface wetting is most likely to be in the transitional region between the Cassie-Baxter and Wenzel states (Fig. 5C). The hydrophobicity (CA) of this sample and *C. atrata* were similar, however the adhesion values on *C. atrata* were significantly higher (over 20%). The patterning on both samples showed similar structure height, however the density of structures was higher on *C. atrata* (Table 1). So it is likely that the increased contact area (and thus enhancement of the van der Waals› forces) between the cicada wing surface and water leads to the high adhesion. The long slender nature of the protrusions on these samples would require significant energy for the water to invade completely to the bottom of the protrusions. The measured CAs for *C. atrata* are lower than the calculated angles (Table 1) based on Cassie-Baxter predictions (and $\cos\theta_w < -1$) (Table S1), so this cuticle is also likely to be in the transitional region (Fig. 5D).

## CONCLUSIONS

We have explored the relationship between structure, wettability and adhesion of several cicada wings. Homogeneity of protrusion patterning on the samples strengthened the hydrophobic properties of the surfaces. As well protusion shape (e.g., nanostructrues with a swollen apex) and density of nanostructuring also play a role in determining adhesion and hydrophobicity. It is also likely that the enhancement of van der Waals forces between water and cicada wings (as a consequence of increased surface wetting) result in a greater force of adhesion. The schematic wetting models of surface nanostructures explain why cicada wings may be highly hydrophobic but not superhydrophobic.

Cicada wing surfaces have been shown in previous studies to exhibit interesting features such as high Young's modulus values (3.7 GPa [27]) and heat resistance (200°C [33]), and have a broad range of applications (e.g. bio-templating) [33]–[35]. We have shown that slight changes in morphologies can achieve the purpose of tunable wettability and adhesion. It is expected that a surface with a sufficiently high adhesive force to a liquid will have many potential applications, such as in liquid transportation and capture without loss and in the analysis of very small volumes of liquid samples.

## MATERIALS AND METHODS

### Ethics Statement

No specific permits were required for the described field studies. The insect species collected are not endangered or protected.

### Preparation of samples

Four different species of cicadas were investigated in this study: *Cryptotympana atrata* (fresh and dry samples), *Meimuna opalifer* (dry sample), *Aola bindusara* (dry sample) and *Leptopsalta bifuscata* (dry sample) collected in Beijing, Shaanxi, Yunnan and Hebei of China, respectively, on different dates (Table 1). The fresh forewing of *C. atrata* was used immediately after collection in order to study age effects on the microstructure, wetting characterization and adhesive forces. The experiments were duplicated on the forewing of *C. atrata* (after naturally drying for more than three months) and the other three species. The apical 1/3 of the forewing of each cicada species was excised with a scalpel, rinsed using deionized water to remove environmental contaminants and then dried at room temperature for measurements on the remigium region.

Microstructure observation

The cicada forewings were attached onto a platform using conductive adhesive. The nanostructures were observed and evaluated on an environmental scanning electron microscope (ESEM) (Quanta 200 FEG, FEI, Eindhoven, Netherlands) after being coated with a thin layer of gold. The wings were titled to 0° and 30°, respectively, for the three dimensional observations. The calculated structure parameters, including basal diameter (*d*), basal spacing (*s*) and height (*h*) of protrusions on the surfaces were an average value of twenty measurements, and their standard deviations (in brackets) were calculated (Table 1).

## Measurements of CA, sliding angle (SA) and CAH

Static CAs and SAs were measured on a Data-Physics OCA 20 contact angle system (Filderstadt, Germany) at ambient temperature using the sessile drop method. The wings were fixed on glass slides with double-sided adhesive. 3 μL water droplets were used for CA measurements and 10 μL water droplets added to wing surfaces via a syringe were used for SA measurements. CAH is defined as the difference between the advancing and receding angles, which were recorded by adding or removing a small amount of water from the drop. Each wetting angle was measured at five to ten different points on each wing, and average values and standard deviations (in brackets) were calculated (Table 1).

## Measurements of adhesive force

The force required to retract the water droplet away from the cicada wing was measured using a high-sensitivity microelectromechanical balance system (Data-Physics DCAT 11, Germany). A 5 μL water droplet was suspended with a metal ring in the first instance, and the cicada wing was placed on the balance table. The cicada wing was moved upward at a constant speed of 0.05 mm $s^{-1}$ until contact with the water droplet. The force increased gradually until it reached its maximum, and the shape of the water droplet changed from spherical to elliptical. When the cicada wing was moved down, the contact force sharply reduced to zero and the shape of the water droplet reverted back to spherical. At this time the maximum force is the adhesion force of water with cicada wing.

## Author Contributions

Conceived and designed the experiments: MXS YMZ. Performed the experiments: MXS JJ. Analyzed the data: MXS APL GSW JAW LJ. Wrote the paper: MXS APL GSW YMZ.

## REFERENCES


1. Barthlott W, Neithhuis C (1997) Purity of the sacred lotus, or escape from contamination in biological surfaces. Planta 202: 1–8.
2. Neithhuis C, Barthlott W (1997) Characterization and distribution of water-repellent, self-cleaning plant surfaces. Ann Bot 79: 667–677.
3. Koch K, Bhushan B, Jung YC, Barthlott W (2009) Fabrication of artificial Lotus leaves and significance of hierarchical structure for superhydrophobicity and low adhesion. Soft Matter 5: 1386–1393.
4. Zhang J, Wang J, Zhao Y, Xu L, Gao X, et al. (2008) How does the leaf margin make the lotus surface dry as the lotus leaf floats on water? Soft Matter 4: 2232–2237.
5. Feng L, Zhang Y, Xi J, Zhu Y, Wang N, et al. (2008) Petal effect: a superhydrophobic state with high adhesive force. Langmuir 24: 4114–4119.
6. Bhushan B, Her EK (2010) Fabrication of superhydrophobic surfaces with high and low adhesion inspired from rose petal. Langmuir 26: 8207–8217.
7. Chang FM, Hong SJ, Sheng YJ, Tsao HK (2009) High contact angle hysteresis of superhydrophobic surfaces: hydrophobic defects. Appl Phys Lett 95: 064102.
8. Joanny JF, de Gennes PG (1984) A model for contact angle hysteresis. J Chem Phys 81: 552–562.
9. Bhushan B (2002) Adhesion. In: Bhushan B, editor. Introduction to tribology. New York: Wiley. pp. 169–206.
10. Jin M, Feng X, Feng L, Sun T, Zhai J, et al. (2005) Superhydrophobic

aligned polystyrene nanotube films with high adhesive force. Adv Mater 17: 1977–1981.

11. Cheng Z, Gao J, Jiang L (2010) Tip geometry controls adhesive states of superhydrophobic surfaces. Langmuir 26: 8233–8238.
12. Zhao W, Wang L, Xue Q (2010) Fabrication of low and high adhesion hydrophobic Au surfaces with micro/nano-biomimetic structures. J Phys Chem C 114: 11509–11514.
13. Cho WK, Choi IS (2008) Fabrication of hairy polymeric films inspired by geckos: wetting and high adhesion properties. Adv Funct Mater 18: 1089–1096.
14. Vincent JFV, Wegst UGK (2004) Design and mechanical properties of insect cuticle. Arthropod Struct Dev 33: 187–199.
15. Holdgate MW (1955) The wetting of insect cuticles by water. J Exp Biol 32: 591–617.
16. Wagner P, Neinhuis C, Barthlott W (1996) Wettability and contaminability of insect wings as a function of their surface sculptures. Acta Zool (Stockholm) 77: 213–225.
17. Gorb SN, Kesel A, Berger J (2000) Microsculpture of the wing surface in Odonata: evidence for cuticular wax covering. Arthropod Struct Dev 29: 129–135.
18. Parker AR, Lawrence CR (2001) Water capture by a desert beetle. Nature 414: 33–34.
19. Cong Q, Chen GH, Fang Y, Ren LQ (2004) Super-hydrophobic characteristics of butterfly wing surface. J Bionics Eng 1: 249–255.
20. Gao X, Jiang L (2004) Water-repellent legs of water striders. Nature 432: 36.
21. Feng XQ, Gao X, Wu Z, Jiang L, Zheng QS (2007) Superior water repellency of water strider legs with hierarchical structures: experiments and analysis. Langmuir 23: 4892–4896.
22. Zheng Y, Gao X, Jiang L (2007) Directional adhesion of superhydrophobic butterfly wings. Soft Matter 3: 178–182.
23. Watson GS, Cribb BW, Watson JA (2010) How micro/nanoarchitecture facilitates anti-wetting: an elegant hierarchical design on the termite wing. ACS Nano 4: 129–136.
24. Watson GS, Cribb BW, Watson JA (2010) The role of micro/nano channel structuring in repelling water on cuticle arrays of the lacewing. J Struc Biol 171: 44–51.
25. Watson JA, Cribb BW, Hu HM, Watson GS (2011) A dual layer hair array of the brown lacewing: repelling water at different length scales.

Biophys J 100: 1149–1155.

26. Hu HM, Watson GS, Cribb BW, Watson JA (2011) Non-wetting wings and legs of the cranefly aided by fine structures of the cuticle. J Exp Biol 214: 915–920.
27. Song F, Lee KL, Soh AK, Zhu F, Bai YL (2005) Experimental studies of the materials properties of the forewing of cicada (Homoptera, Cicadidae). J Exp Biol 207: 3035–3042.
28. Stoddart PR, Cadusch PJ, Boyce TM, Erasmus RM, Comins JD (2006) Optical properties of chitin: surface-enhanced Raman scattering substrates based on antireflection structures on cicada wings. Nanotechnology 17: 680–686.
29. Zhang G, Zhang J, Xie G, Liu Z, Shao H (2006) Cicada wings: a stamp from nature for nanoimprint lithography. Small 2: 1440–1443.
30. Watson JA, Myhra S, Watson GS (2006) Tunable natural nano-arrays: controlling surface properties and light reflectance. Proc SPIE 6037: 375–383.
31. Watson GS, Myhra S, Cribb BW, Watson JA (2008) Putative functions and functional efficiency of ordered cuticular nanoarrays on insect wings. Biophys J 94: 3352–3360.
32. Xie G, Zhang G, Lin F, Zhang J, Liu Z, et al. (2008) The fabrication of subwavelength anti-reflective nanostructures using a bio-template. Nanotechnology 19: 095605.
33. Hong SH, Hwang J, Lee H (2009) Replication of cicada wing's nano-patterns by hot embossing and UV nanoimprinting. Nanotechnology 20: 385303.
34. Kostovski G, White DJ, Mitchell A, Austin MW, Stoddart PR (2009) Nanoimprinted optical fibres: biotemplated nanostructures for SERS sensing. Biosens Bioelectron 24: 1531–1535.
35. Kostovski G, Chinnasamy U, Jayawardhana S, Stoddart PR, Mitchell A (2011) Sub-15nm optical fiber nanoimprint lithography: a parallel, self-aligned and portable approach. Adv Mater 23: 531–535.
36. Sun MX, Liang AP, Zheng YM, Watson GS, Watson JA (2011) A study of the antireflection efficiency of natural nano-arrays of varying sizes. Bioinspir Biomim 6: 026003.
37. Hu HM, Watson JA, Cribb BW, Watson GS (2011) Fouling of nanostructured insect cuticle: adhesion of natural and artificial contaminants. Biofouling 27: 1125–1137.
38. Green DW, Watson GS, Watson J, Abraham SJK (2012) New biomimetic directions in regenerative ophthalmology. Adv Healthcare Mater. DOI:10.1002/adhm.201100039.

39. Sun MX, Watson GS, Zheng YM, Watson JA, Liang AP (2009) Wetting properties on nanostructured surfaces of cicada wings. J Exp Biol 212: 3148–3155.

40. Quéré D, Reyssat M (2008) Non-adhesive lotus and other hydrophobic materials. Phil Trans R Soc A 366: 1539–1556.

41. Wang S, Jiang L (2007) Definition of superhydrophobic states. Adv Mater 19: 3423–3424.

42. Wenzel RN (1936) Resistance of solid surfaces to wetting by water. Ind Eng Chem Res 28: 988–994.

43. Cassie ABD, Baxter S (1944) Wettability of porous surfaces. Trans Faraday Soc 4: 546–551.

44. Bico J, Thiele U, Quéré D (2002) Wetting of textured surfaces. Colloids Surf A 206: 41–46.

# Citations

## CHAPTER 1

Dehzangi A, Larki F, Hutagalung SD, Goodarz Naseri M, Majlis BY, et al. (2013) Impact of Parameter Variation in Fabrication of Nanostructure by Atomic Force Microscopy Nanolithography. PLoS ONE 8(6): e65409. doi:10.1371/journal.pone.0065409

## CHAPTER 2

Halvorsen K, Wong WP (2012) Binary DNA Nanostructures for Data Encryption. PLoS ONE 7(9): e44212. doi:10.1371/journal.pone.0044212

## CHAPTER 3

Silva JM, Georgi N, Costa R, Sher P, Reis RL, et al. (2013) Nanostructured 3D Constructs Based on Chitosan and Chondroitin Sulphate Multilayers for Cartilage Tissue Engineering. PLoS ONE 8(2): e55451. doi:10.1371/journal.pone.0055451

## CHAPTER 4

Rahman MM, Khan SB, Asiri AM (2014) Fabrication of Smart Chemical Sensors Based on Transition-Doped-Semiconductor Nanostructure Materials with µ-Chips. PLoS ONE 9(1): e85036. doi:10.1371/journal.pone.0085036

## CHAPTER 5

Sergi PN, Morana Roccasalvo I, Tonazzini I, Cecchini M, Micera S (2013) Cell Guidance on Nanogratings: A Computational Model of the Interplay between PC12 Growth Cones and Nanostructures. PLoS ONE 8(8): e70304. doi:10.1371/journal.pone.0070304

## CHAPTER 6

Xu J, Li Y, Xiang Y, Chen X (2013) A Super Energy Mitigation Nanostructure at High Impact Speed Based on Buckyball System. PLoS ONE 8(5): e64697. doi:10.1371/journal.pone.0064697

## CHAPTER 7

Brasen JC, Jacobsen JCB, Holstein-Rathlou N-H (2012) The Nanostructure of Myoendothelial Junctions Contributes to Signal Rectification between Endothelial and Vascular Smooth Muscle Cells. PLoS ONE 7(4): e33632. doi:10.1371/journal.pone.0033632

## CHAPTER 8

Wolun-Cholewa M, Langer K, Szymanowski K, Glodek A, Jankowska A, et al. (2013) An Efficient 3D Cell Culture Method on Biomimetic Nanostructured Grids. PLoS ONE 8(9): e72936. doi:10.1371/journal.pone.0072936

## CHAPTER 9

Zhou D, Ji Z, Jiang X, Dunphy DR, Brinker J, et al. (2013) Influence of Material Properties on TiO2 Nanoparticle Agglomeration. PLoS ONE 8(11): e81239. doi:10.1371/journal.pone.0081239

## CHAPTER 10

Forsgren J, Frykstrand S, Grandfield K, Mihranyan A, Strømme M (2013) A Template-Free, Ultra-Adsorbing, High Surface Area Carbonate Nanostructure. PLoS ONE 8(7): e68486. doi:10.1371/journal.pone.0068486

## CHAPTER 11

Sun M, Liang A, Watson GS, Watson JA, Zheng Y, et al. (2012) Influence of Cuticle Nanostructuring on the Wetting Behaviour/States on Cicada Wings. PLoS ONE 7(4): e35056. doi:10.1371/journal.pone.0035056

# INDEX

## E



## F



## G



## H



## I



## J



## K



## L



## M

**V**



**W**



**X**



**Y**